Lucio Boccardo, Gisella Croce

Elliptic Partial Differential Equations

De Gruyter Studies in Mathematics

Volume 55

Lucio Boccardo, Gisella Croce

Elliptic Partial Differential Equations

Existence and Regularity of Distributional Solutions

DE GRUYTER

Mathematics Subject Classification 2010
35J20, 35J25, 35J60, 35J61, 35J62, 35J70, 49J45, 49R05

Authors
Prof. Lucio Boccardo
Sapienza Università di Roma
Dipartimento di Matematica
Piazzale Aldo Moro 5
00185 Roma
ITALY
boccardo@mat.uniroma1.it

Prof. Gisella Croce
Laboratoire de Mathématiques Appliquées du Havre
Université du Havre
25, rue Philippe Lebon
76058 Le Havre
FRANCE
croceg@univ-lehavre.fr

ISBN 978-3-11-031540-0
e-ISBN 978-3-11-031542-4
Set-ISBN 978-3-11-031543-1
ISSN 0179-0986

Library of Congress Cataloging-in-Publication Data
A CIP catalog record for this book has been applied for at the Library of Congress.

Bibliographic information published by the Deutsche Nationalbibliothek
The Deutsche Nationalbibliothek lists this publication in the Deutsche Nationalbibliografie; detailed bibliographic data are available in the Internet at http://dnb.dnb.de.

Typesetting: le-tex publishing services GmbH, Leipzig
Printing and binding: Hubert & Co. GmbH & Co. KG, Göttingen
♾ Printed on acid-free paper
Printed in Germany

www.degruyter.com

Bernardo m'accennava, e sorridea,
perch'io guardassi suso …
(Paradiso, XXXIII, vv. 49–50)

This book is dedicated to Bernard Dacorogna for his 60th birthday.

Contents

Notations

Ω	open bounded set of $\mathbb{R}^N$
$\partial\Omega$	boundary of Ω
meas	Lebesgue measure in $\mathbb{R}^N$
a.e.	almost everywhere (with respect to the Lebesgue measure)
X'	dual of X (i.e., space of linear and continuous functionals on X) where X is a Banach space
X''	dual of X' where X is a Banach space
$\mathcal{L}(X)$	space of continuous and linear operators from a Banach space X into X
$\langle \varphi, v \rangle$	$\varphi(v)$ if $\varphi \in X'$ and $v \in X$
$L^p(\Omega)$	space of functions f such that $\lvert f\rvert^p$ is Lebesgue integrable over $\Omega \subset \mathbb{R}^N$
$C^k(\Omega)$	space of k times differentiable functions on Ω with continuity
$C_0^k(\Omega)$	space of k times differentiable functions on Ω, with continuity, 0 on $\partial\Omega$
$C^1(\Omega, \mathbb{R}^N)$	space of differentiable maps from Ω onto $\mathbb{R}^N$ with continuity
$C^1(\mathbb{R}^N, \mathbb{R}^N)$	space of differentiable maps from $\mathbb{R}^N$ onto $\mathbb{R}^N$ with continuity
$W^{1,p}(\Omega)$	(Sobolev) space of functions with distributional gradient in $(L^p(\Omega))^N$
$W_0^{1,p}(\Omega)$	closure of $C^1(\Omega)$ functions with compact support in Ω with respect to the $W^{1,p}(\Omega)$ norm
$H_0^1(\Omega)$	$W_0^{1,2}(\Omega)$
$H^{-1}(\Omega)$	$(H_0^1(\Omega))'$
$W^{-1,p'}(\Omega)$	$(W_0^{1,p}(\Omega))'$
$W^{m,2}(\Omega)$	space of $W^{m-1,2}(\Omega)$ functions, whose distributional gradient is in $W^{m-1,2}(\Omega)$
$H^m(\Omega)$	$W^{m,2}(\Omega)$
p'	$\frac{p}{p-1}$
p^*	$\frac{Np}{N-p}$
$\chi_E(x)$	$\begin{cases} 1, & \text{if } x \in E \\ 0, & \text{elsewhere} \end{cases}$ for a given set $E \subset \mathbb{R}^N$
$\{u \geq 0\}$	$\{x \in \mathbb{R}^N : u(x) \geq 0\}$ for a given function u
for a given function u,	$u^+ = u\chi_{\{u\geq 0\}}$; $u^- = -u\chi_{\{u\leq 0\}}$
$T_k(x)$	$\begin{cases} -k, & x \leq -k \\ x, & \lvert x\rvert \leq k \\ k, & x \geq k \end{cases}$ for $k > 0$
$G_k(x)$	$x - T_k(x)$ for $k > 0$
$B(x_0, r)$	$\{x \in \mathbb{R}^N : \lvert x - x_0\rvert < r\}$ for a given $x_0 \in \mathbb{R}^N$ and $r > 0$
supp	support of a function
$\mathrm{sgn}(x)$	sign of x for $x \neq 0$, that is, $\frac{x}{\lvert x\rvert}$

1 Introduction

This book is the result of undergraduate and Ph.D. courses that the first author has given at La Sapienza University of Rome on elliptic problems. With the aim of giving the background to those who want to do research in this mathematical field, we have divided this book into two parts. In the first one we present some classical results about the existence and the regularity of weak solutions to elliptic problems in divergence form (so that we do not discuss the theory of classical solutions). After the semilinear equations, we study the Leray–Lions problem

$$\begin{cases} -\mathrm{div}(a(x,u,\nabla u)) = f\,, & \text{in}\,\Omega\,,\\ u = 0\,, & \text{on}\,\partial\Omega\,, \end{cases}$$

where a is an elliptic operator, that is, $a(x,s,\xi)\cdot\xi \geq \alpha|\xi|^2$, and f belongs to $H^{-1}(\Omega)$. We prove the existence and regularity results due to Leray–Lions and Stampacchia. We also treat the spectral theory of linear operators and the H^2 regularity of solutions to linear problems. Even if in this book we focus our attention on partial differential equations, we have presented a chapter about the Calculus of Variations, since some of the differential problems that we treat are motivated by the minimization of integral functionals.

The Leray–Lions problem is the origin of an important and still active area of research. In the second part of this book we exhibit some possible directions: the existence of solutions when the source f has a low summability (for instance f is $L^1(\Omega)$ function or it is a measure) and the uniqueness of solutions. Moreover we will study three problems where the hypotheses of the Leray–Lions problems are relaxed. Indeed we will present a problem defined by an elliptic operator with a quadratic term in the gradient; we will also study an equation defined by an elliptic operator with a polynomial growth term. Finally we will analyze a problem defined by an elliptic operator with degenerate coercivity.

We have made an effort to keep this book self-contained. However, for background in real analysis, functional analysis, and Sobolev spaces, we refer to the book *Functional analysis, Sobolev spaces and partial differential equations* of Haïm Brezis [22]. For the reader's convenience we have collected in the Appendices all the main prerequisites.

Lucio Boccardo and Gisella Croce

Part I

2 Some fixed point theorems

2.1 Introduction

In the study of the existence and uniqueness of solutions to differential equations the fixed point theorems have a very important role. Indeed, it is often possible to look for the solutions to a differential problem among the fixed points of a certain operator related to the differential problem. We will present three fixed point theorems in this chapter.

The first result, the Banach–Caccioppoli theorem, states that a function from a complete metric space to the same space which contracts the distances has a unique fixed point.

Thereafter, we will prove Brouwer's theorem, which assures the existence of a fixed point for a continuous map from a convex bounded closed subset of $\mathbb{R}^N$ to itself.

We will finally present Schauder's theorem, which is a useful tool in the study of certain differential problems, as we will see. In some sense, it is the analog of Brouwer's theorem for operators defined on any Banach spaces.

2.2 Banach–Caccioppoli theorem

Theorem 2.1. *Let (X, d) be a complete metric space and let $F : X \to X$ be a map with the following property: there exists $\theta \in (0, 1)$ such that*

$$d(F(x), F(y)) \leq \theta d(x, y), \quad \forall x, y \in X. \tag{2.2.1}$$

Then there exists a unique $x \in X$ such that $F(x) = x$, i.e., a unique fixed point of F.

Remark 2.2. A map F that satisfies condition (2.2.1) is called contraction.

Remark 2.3. Among the fixed point theorems that we present in this chapter, only the Banach–Caccioppoli theorem gives the uniqueness. We recall that it is used in the study of Cauchy problems for ordinary differential equations.

Proof. The proof is based on an elementary iteration argument. Let us fix any $x_0 \in X$ and let us define

$$x_n := F(x_{n-1}), \quad n \geq 1. \tag{2.2.2}$$

Using hypothesis (2.2.1), we have

$$d(x_{n+1}, x_n) = d\left(F(x_n), F(x_{n-1})\right) \leq \theta d(x_n, x_{n-1}) \leq \cdots \leq \theta^n d(x_1, x_0) \tag{2.2.3}$$

for every $n \geq 0$. The triangle inequality and (2.2.3) give, for every $p \in \mathbb{N}$,

$$d(x_{n+p+1}, x_n) \leq \sum_{i=1}^{p+1} d(x_{n+i}, x_{n+i-1}) \leq \left(\theta^{n+p} + \cdots + \theta^n\right) d(x_1, x_0).$$

Now, applying Cauchy's criterion for series to $\sum_{n=1}^{\infty} \theta^n$, we deduce that x_n is a Cauchy series. Since X is complete, x_n converges to some $x \in X$. The continuity of F, given by (2.2.1), implies that $F(x_n)$ converges to $F(x)$. Passing to the limit in (2.2.2), we get the existence of a fixed point.

For the uniqueness, let x, y be two fixed points of F. By hypothesis (2.2.1)

$$d(x,y) = d(F(x),F(y)) \leq \theta\, d(x,y)\,.$$

Since $\theta < 1$ we deduce that $x = y$, that is, there exists one and only one fixed point. □

2.3 Brouwer's theorem

Theorem 2.4 (Brouwer). *Let K be a convex, closed, and bounded subset of $\mathbb{R}^N$ and let $f : K \to K$ be a continuous function. Then f has a fixed point.*

Remark 2.5. The hypotheses on f are different from those of Theorem 2.1: indeed, in Brouwer's theorem f is only assumed to be continuous; however the existence of a convex, closed, and bounded invariant set is required.

We will follow the proof given in [35] (note that there exist several different proofs: for example there is one which uses the notion of topological degree and another one uses the notion of the homology group). Observe that we will make use of Banach–Caccioppoli theorem.

We will denote with $B(0,r)$ the set $\{x \in \mathbb{R}^N : |x| < r\}$, for $r > 0$.

Theorem 2.6. *Let $F : \overline{B(0,1)} \to \partial B(0,1)$ be a continuous function. Then there exists $x \in \partial B(0,1)$ such that $F(x) \neq x$.*

Proof. We assume by contradiction that $F(x) = x$ for every $x \in \partial B(0,1)$. Let us define the following continuous extension of F:

$$\tilde{f}(x) = \begin{cases} F(x)\,, & \text{if } |x| \leq 1\,, \\ \dfrac{x}{|x|}\,, & \text{if } |x| > 1; \end{cases}$$

we remark that $|\tilde{f}(x)| = 1$. The Weierstrass density theorem implies the existence of a $C^1(\mathbb{R}^N, \mathbb{R}^N)$ map f_1 such that

$$\sup_{x \in \overline{B(0,2)}} \left|\tilde{f}(x) - f_1(x)\right| < \frac{1}{2}. \tag{2.3.1}$$

Let us consider any $\phi \in C^1(\mathbb{R})$ such that $0 \leq \phi \leq 1$ and

$$\phi(t) = \begin{cases} 1\,, & \text{if } t \leq 3/2\,, \\ 0\,, & \text{if } t \geq 2\,, \end{cases}$$

with ϕ decreasing for $t \in (3/2, 2)$. We define the following combination of $\tilde{f}$ and f_1:

$$f_c(x) = [1 - \phi(|x|)]\tilde{f}(x) + \phi(|x|)\, f_1(x)\,.$$

Finally, we set

$$N(x) = \frac{f_c(2x)}{|f_c(2x)|}\,.$$

With these definitions in mind, we prove the theorem in three steps.

Step I: Let us prove that N is $C^1(\mathbb{R}^N, \mathbb{R}^N)$ and Lipschitz continuous. For the regularity it suffices to prove that $f_c \in C^1(\mathbb{R}^N, \mathbb{R}^N)$ and that $f_c(x) \neq 0$ for every $x \in \mathbb{R}^N$. Now, we observe that if $|x| > 1$

$$f_c(x) = (1 - \phi(|x|))\,\frac{x}{|x|} + \phi(|x|)\, f_1(x)$$

and so f_c is a C^1 function in the set $\{x \in \mathbb{R}^N : |x| > 1\}$. On the other hand, if $|x| < 3/2$ then $f_c = f_1$, and $f_1 \in C^1(\mathbb{R}^N, \mathbb{R}^N)$ by definition; therefore $f_c \in C^1(\mathbb{R}^N, \mathbb{R}^N)$. Also $f_c \neq 0$ for every $x \in \mathbb{R}^N$. Indeed

$$|f_c(x)| \geq \left|\tilde{f}(x)\right| - \phi(|x|)\left|\tilde{f}(x) - f_1(x)\right| \geq 1 - \left|\tilde{f}(x) - f_1(x)\right| > \tfrac{1}{2}\,,$$

due to (2.3.1). Thus $N \in C^1(\mathbb{R}^N, \mathbb{R}^N)$.

Moreover N is Lipschitz continuous in $\overline{B(0,1)}$, since $N \in C^1(\overline{B(0,1)}, \mathbb{R}^N)$. Away from $\overline{B(0,1)}$, it holds that

$$f_c(2x) = [1 - \phi(2|x|)]\tilde{f}(2x) + \phi(2|x|) f_1(2x) = \tilde{f}(2x) = \frac{x}{|x|}\,.$$

Therefore, away from $\overline{B(0,1)}$

$$N(x) = \frac{x}{|x|}$$

which is clearly Lipschitz continuous if $|x| > 1$. Consequently there exists $M > 0$ such that

$$|N(v) - N(w)| \leq M\,|v - w|\,, \quad \forall\, v, w \in \mathbb{R}^N\,. \tag{2.3.2}$$

Step II: Let us prove that $I + t\,N$ is a diffeomorphism, for $t \in (0, \frac{1}{M})$, from $\overline{B(0,1)}$ to $\overline{B(0, t+1)}$. It is easily seen that the image of $\overline{B(0,1)}$ is contained in $\overline{B(0, t+1)}$, since, if $|x| \leq 1$, then $|x + t\,N(x)| \leq |x| + t\,|N(x)| \leq 1 + t$. Moreover, if $y \in \overline{B(0, t+1)}$, there exists a unique $x \in \overline{B(0,1)}$ such that $y = x + tN(x)$. Indeed the map $T : \mathbb{R}^N \to \mathbb{R}^N$, defined by $T(x) = y - t\,N(x)$, is a contraction by our choice of t, since

$$|T(v) - T(w)| = t\,|N(v) - N(w)| \leq t\,M\,|v - w|\,, \quad \forall\, v, w \in \mathbb{R}^N\,,$$

by (2.3.2). Theorem 2.1 implies the existence of a unique $x_0 = y - t\,N(x_0)$. Let us prove that $|x_0| \leq 1$. Assume by contradiction that $|x_0| > 1$; we have

$$|y| = \left| x_0 + t\frac{x_0}{|x_0|} \right| = |x_0| + t > 1 + t\,.$$

This is absurd, since $y \in \overline{B(0, t+1)}$. It remains to prove that

$$\det(D(I + tN)) \neq 0$$

where D denotes the gradient of the map. If there were $y \neq 0$ such that $y = -t(DN)y$, we would have $|y| \leq t|DN||y| \leq t\,M|y|$; this is in contradiction with the choice of t. Thus $I + t\,N$ is a diffeomorphism between $\overline{B(0,1)}$ and $\overline{B(0, t+1)}$.

Step III: Let us prove the existence of $x_0 \in B(0,1)$ such that $\det DN(x_0) \neq 0$. Let us distinguish the cases $\det(I + t\,DN) > 0$ and $\det(I + t\,DN) < 0$. In the first case, using a change of variables we have

$$(1+t)^N \int\limits_{B(0,1)} dy = \int\limits_{B(0,t+1)} dy = \int\limits_{B(0,1)} \det(I + tDN(x))\, dx\ .$$

The last term is a polynomial in t of degree N, whose leading coefficient is

$$\int\limits_{B(0,1)} \det(DN(x))dx\ .$$

The identity principle of polynomials implies that

$$\int\limits_{B(0,1)} dy = \int\limits_{B(0,1)} \det(DN(x))dx\,.$$

From this identity we infer the existence of $x_0 \in B(0,1)$ such that $\det DN(x_0) \neq 0$. In the case $\det(I + DN) < 0$, with a similar argument, one has

$$(1+t)^N \int\limits_{B(0,1)} dy = \int\limits_{B(0,t+1)} dy = -\int\limits_{B(0,1)} \det(I + tDN(x))\, dx.$$

As before,

$$\int\limits_{B(0,1)} dy = -\int\limits_{B(0,1)} \det(DN(x))dx$$

and so there exists $x_0 \in B(0,1)$ such that $\det DN(x_0) \neq 0$.

Now, since DN is an isomorphism, its kernel is composed of only the zero vector. $N(x_0)$ belongs to the kernel of $DN(x_0)$: indeed, from the identity

$$(N(x)|N(x)) = 1\,,$$

we deduce that $DN(x)\,N(x) = 0$ for every $x \in \mathbb{R}^N$ and so $N(x_0) = 0$. This is a contradiction, as $|N(x)| = 1$ for every $x \in \mathbb{R}^N$. □

We can now prove Brouwer's theorem.

Proof. We divide the proof into two steps.

Step I: Let us prove the result in the case where $K = \overline{B(0,1)}$. By contradiction, assume that $f(x) \neq x$, for every $x \in \overline{B(0,1)}$. Define $F(x)$, for every $x \in \overline{B(0,1)}$, as the intersection point between the half-line $f(x)+\lambda(x-f(x))$, $\lambda \geq 0$, and $\partial B(0,1)$. We claim that F is continuous. We have, for some $t(x) \in (0,1]$

$$x = t(x)F(x) + (1-t(x))f(x)\,. \tag{2.3.3}$$

This yields

$$F(x) = s(x)\,x + (1-s(x))\,f(x)\,,$$

where $s(x) = \frac{1}{t(x)} \geq 1$ is such that $|F(x)| = 1$. We are going to prove that $s(x)$ is continuous; this will imply that F is continuous. One has, since $F(x) \in \partial B(0,1)$

$$1 = |F(x)|^2 = s^2(x)\,|x-f(x)|^2 + |f(x)|^2 + 2s(x)\,(x-f(x)|f(x))\,,$$

that is,

$$s^2(x)\,|x-f(x)|^2 + 2s(x)\,(x-f(x)|f(x)) + |f(x)|^2 - 1 = 0\,.$$

Now, for a fixed x,

$$\psi(s) = s^2|x-f(x)|^2 + 2s(x-f(x)|f(x)) + |f(x)|^2 - 1$$

is a polynomial of second degree, and so it has at most two zeros. Since $\psi(1) \leq 0$ and $\lim_{s\to\infty}\psi(s) = +\infty$, ψ has only one zero in $[1,+\infty)$. This implies that $s(x)$ is well defined. Moreover, s is continuous, since it is a zero of a polynomial of second degree with continuous coefficients.

Let us prove that $F(x) = x$, for $|x| = 1$. From (2.3.3) we have to prove that $t(x) = 1$. If $t(x) \neq 1$, squaring (2.3.3) we obtain

$$t^2(x) + (1-t(x))^2\,|f(x)|^2 + 2t(x)\,(1-t(x))\,(F(x)|f(x)) = 1\,,$$

that is,

$$(1-t(x))^2\,|f(x)|^2 + 2t(x)\,(1-t(x))\,(F(x)|f(x)) = (1-t(x))\,(1+t(x))\,.$$

Dividing by $1-t(x)$ we get

$$t(x)\,[2(F(x)|f(x)) - 1 - |f(x)|^2] = 1 - |f(x)|^2\,,$$

that is,

$$t(x)\,|F(x)-f(x)|^2 = -1 + |f(x)|^2\,.$$

This is a contradiction, since $|f|^2 - 1 < 0$. Therefore $t(x) = 1$ and $x = F(x)$. We have thus defined a continuous map $F : \overline{B(0,1)} \to \partial B(0,1)$ and all the points of $\partial B(0,1)$ are fixed points: this is in contradiction with Theorem 2.6.

Step II: Let us study the case where K is any convex compact set. Since K is bounded, there exists $R > 0$ such that $K \subset \overline{B(0,R)}$. Let P_K be the projection over K and

$$\begin{aligned} \tilde{f}: \quad \overline{B(0,R)} &\to K \subset \overline{B(0,R)} \\ x &\to f(P_K(x)) . \end{aligned}$$

Due to Step I, $\tilde{f}$ has a fixed point $x \in \overline{B(0,R)}$; the image of $\tilde{f}$ is in K, so $x \in K$. □

Remark 2.7. In dimension 1, Brouwer's theorem states that if $f : [a,b] \to [a,b]$ is continuous, then f has a fixed point. In this case, the existence of a fixed point can be proved in a very simple way: it is sufficient to apply the intermediate value theorem to $\psi(t) = t - f(t)$.

2.4 Schauder's theorem

Schauder's theorem is a fixed point result for operators over Banach spaces. We will present two different versions (see [46]).

Theorem 2.8. *Assume that X is a Banach space. Let K be a compact convex subset of X invariant under a continuous map $F : K \subset X \to X$. Then F has a fixed point in K.*

To state the second version of Schauder's theorem, we need the following definition:

Definition 2.9. A map $T : X \to X$ is completely continuous if it is continuous and if, for every bounded subset B of X, $\overline{T(B)}$ is compact.

Theorem 2.10. *Let F be a completely continuous map and let K be a convex, bounded, closed and invariant subset of X. Then F has a fixed point in K.*

We are going to prove Theorem 2.8. We observe that we will make use of Brouwer's theorem. $\|x\|$ will denote the norm of an element $x \in X$ and $B(y,r) = \{x \in X : \|x - y\| < r\}$.

Proof (of Theorem 2.8). Let $\varepsilon > 0$. Since K is compact, there exist $x_1, \ldots, x_{N_\varepsilon} \in K$ such that

$$K \subset \bigcup_{i=1}^{N_\varepsilon} B(x_i, \varepsilon) . \tag{2.4.1}$$

Now, let E_ε be the vector space generated by $\{x_1, \ldots, x_{N_\varepsilon}\}$ and let $b_j : K \to \mathbb{R}$, $j = 1, \ldots, N_\varepsilon$, be defined by

$$b_j(x) = (\varepsilon - \|x - x_j\|)^+ .$$

Since for every $x \in K$, not every $b_j(x)$ is zero, we can define

$$G_\varepsilon(x) = \frac{\sum_{j=1}^{N_\varepsilon} b_j(x)\, x_j}{\sum_{j=1}^{N_\varepsilon} b_j(x)} .$$

G_ε is a convex combination of $x_1, \dots, x_{n_\varepsilon}$. This implies that $G_\varepsilon(K) \subset E_\varepsilon \cap K$, as K is convex. We remark that G_ε is continuous. We can apply Brouwer's theorem to the function $G_\varepsilon \circ F$ and to the compact convex set $E_\varepsilon \cap K$. Then there exists $x_\varepsilon \in K \cap E_\varepsilon$ such that

$$G_\varepsilon(F(x_\varepsilon)) = x_\varepsilon \,. \tag{2.4.2}$$

Since $K \cap E_\varepsilon$ is compact, there exists a subsequence x_ε converging to some $x_0 \in K$. The continuity of F implies that $F(x_\varepsilon) \to F(x_0)$. On the other hand, observe that for every $x \in K$ it holds

$$\begin{aligned} \|G_\varepsilon(x) - x\| &= \left\| \frac{\sum_{j=1}^{N_\varepsilon} b_j(x) x_j}{\sum_{j=1}^{N_\varepsilon} b_j(x)} - x \right\| = \frac{\left\| \sum_{j=1}^{N_\varepsilon} b_j(x)(x_j - x) \right\|}{\sum_{j=1}^{N_\varepsilon} b_j(x)} \\ &\le \frac{\sum_{j=1}^{N_\varepsilon} b_j(x) \|x_j - x\|}{\sum_{j=1}^{N_\varepsilon} b_j(x)} \le \frac{\sum_{j=1}^{N_\varepsilon} b_j(x) \varepsilon}{\sum_{j=1}^{N_\varepsilon} b_j(x)} = \varepsilon \,, \end{aligned} \tag{2.4.3}$$

by (2.4.1). From (2.4.2) and the fact that $F(x_\varepsilon) \in K$ one has

$$\varepsilon > \|G_\varepsilon(F(x_\varepsilon)) - F(x_\varepsilon)\| = \|x_\varepsilon - F(x_\varepsilon)\| \,.$$

At the limit as $\varepsilon \to 0$, one gets $x_0 = F(x_0)$. □

Proof (of Theorem 2.10). Let us fix $\varepsilon > 0$. Since $\overline{F(K)}$ is compact, there exist $v_1, \dots, v_{N_\varepsilon} \in \overline{F(K)} \subset K$ such that

$$F(K) \subset \bigcup_{i=1}^{N_\varepsilon} B(v_i, \varepsilon) \,. \tag{2.4.4}$$

Now, let E_ε be the vector space generated by $\{v_1, \dots, v_{N_\varepsilon}\}$ and let $b_j : K \to \mathbb{R}$ be defined by

$$b_j(x) = (\varepsilon - \|x - v_j\|)^+ \,.$$

For every $u \in \overline{F(K)}$, we define

$$G_\varepsilon(u) = \frac{\sum_{j=1}^{N_\varepsilon} b_j(u) v_j}{\sum_{j=1}^{N_\varepsilon} b_j(u)} \,.$$

$G_\varepsilon(u)$ is a convex combination of $v_1, \dots, v_{N_\varepsilon}$: this implies that $G_\varepsilon \circ F(K \cap E_\varepsilon) \subset K \cap E_\varepsilon$, as K is convex. We remark that $G_\varepsilon \circ F$ is continuous. Brouwer's theorem gives the existence of $x_\varepsilon \in K \cap E_\varepsilon$ such that $G_\varepsilon(F(x_\varepsilon)) = x_\varepsilon$. We observe that, for $x \in K$, one has

$$\begin{aligned} \|G_\varepsilon(F(x)) - F(x)\| &= \frac{\left\| \sum_{j=1}^{N_\varepsilon} b_j(F(x))(v_j - F(x)) \right\|}{\sum_{j=1}^{N_\varepsilon} b_j(F(x))} \\ &\le \frac{\sum_{j=1}^{N_\varepsilon} b_j(F(x)) \|v_j - F(x)\|}{\sum_{j=1}^{N_\varepsilon} b_j(F(x))} \le \varepsilon \,, \end{aligned}$$

by (2.4.4). This estimate and the fact that $G_\varepsilon(F(x_\varepsilon)) = x_\varepsilon$ yield

$$\|F(x_\varepsilon) - x_\varepsilon\| \to 0\,, \quad \varepsilon \to 0\,.$$

On the other hand, $F(x_\varepsilon) \in \overline{F(K)}$ which is compact; up to a subsequence, $F(x_\varepsilon) \to x_0$. Now,

$$\|x_\varepsilon - x_0\| \le \|x_\varepsilon - F(x_\varepsilon)\| + \|F(x_\varepsilon) - x_0\|$$

and the right-hand side tends to 0, as $\varepsilon \to 0$. This implies that $x_\varepsilon \to x_0$ and $F(x_\varepsilon) \to F(x_0)$ from the continuity of F. We have already proved that $F(x_\varepsilon) \to x_0$. The uniqueness of the limit of $F(x_\varepsilon)$ gives the result. □

Remark 2.11. In Brouwer's theorem, the invariant set is closed and bounded, that is, compact, since it is a set of $\mathbb{R}^N$. In infinite dimensional spaces, due to a theorem by Riesz, the closed and bounded sets are not necessarily compact in general (see [22]). This is why compactness is required in Schauder's theorems.

The following example, due to Kakutani [31], shows that we can define in l^2 a continuous operator which maps the unit ball to itself without admitting a fixed point.

Example 2.12. Let $T : l^2 \to l^2$ be defined by

$$T(x) = \left(\frac{1}{2}(1 - \|x\|^2), x_1, x_2, \dots\right)\,,$$

for every $x = (x_1, x_2, \dots) \in l^2$, where $\|x\|^2 = \sum_{i=1}^{\infty} |x_i|^2$. T is continuous, since

$$\|T(x) - T(y)\|^2 = \frac{1}{4}\left[\|y\|^2 - \|x\|^2\right]^2 + \|x - y\|^2\,.$$

Moreover the unit ball is invariant: indeed, if $\|x\| \le 1$, one has

$$\|T(x)\|^2 = \left[\frac{1}{2}(1 - \|x\|^2)\right]^2 + \|x\|^2 \le 1\,,$$

since $t \to 1 - \frac{1}{4}(1 - t^2)^2 - t^2$ is positive and decreasing for $t \in [0, 1]$. On the other hand, it is easily seen that T has no fixed point in the unit ball. Indeed, if $\|x\| = 1$, we have $T(x) = (0, x_1, x_2, \dots)$; if $T(x) = x$, then $x_j = 0$, for every j, and so $\|x\| = 0 \ne 1$. If $||x|| = \theta < 1$, then $T(x) = (\frac{1}{2}(1 - \theta^2), x_1, x_2, \dots)$; if $T(x) = x$, then $x_j = \frac{1}{2}(1 - \theta^2)$, for every j. This is impossible, since $x \in l^2$.

3 Preliminaries of real analysis

3.1 Introduction

This chapter, devoted to some important results of real analysis, is divided into two parts. In the first one we will present Nemitski's composition theorem. This result establishes the continuity of an operator defined between two Lebesgue spaces, through the composition with a real function. In the second part of this chapter we will define the Marcinkiewicz spaces M^p, since they will be used later in several regularity results.

3.2 Nemitski's composition theorem

The aim of this section is to study the continuity of

$$\Phi: \quad L^p(\Omega) \mapsto L^q(\Omega)$$
$$u(x) \mapsto f(x, u(x))$$

defined between two Lebesgue spaces through the composition with a function f.

The proof of the result that we will present is based on several theorems of real analysis. The following ones give some convergence results in L^p. In the sequel Ω will be an open bounded subset of $\mathbb{R}^N$.

Theorem 3.1. *Let f_n be a sequence of functions and f be a function in $L^p(\Omega)$, $p > 1$. Assume that*

(1) *f_n is bounded in $L^p(\Omega)$;*

(2) *$f_n \to f$ a.e. in Ω.*

Then $f_n \to f$ in $L^q(\Omega)$, for every $q \in [1, p)$ and weakly in $L^p(\Omega)$.

Proof. Assumption (1) implies the existence of a constant $L > 0$ such that

$$\|f_n - f\|_{L^p(\Omega)} \le L, \quad \forall\, n \in \mathbb{N}. \tag{3.2.1}$$

Let $k \in \mathbb{R}^+$. We have

$$\begin{aligned} k^p \operatorname{meas}(\{|f_n - f| > k\}) &\le \int_{\{|f_n-f|>k\}} |f_n - f|^p \\ &\le \int_\Omega |f_n - f|^p \le L^p . \end{aligned} \tag{3.2.2}$$

For every $q \in [1, p)$, one has

$$\int_\Omega |f_n - f|^q = \int_{\{|f_n-f|>k\}} |f_n - f|^q + \int_{\{|f_n-f|\le k\}} |f_n - f|^q .$$

Using Hölder's inequality with exponent $\frac{p}{q}$ on the right-hand side of this equality, we get

$$\int_\Omega |f_n - f|^q \le \left(\int_{\{|f_n-f|>k\}} |f_n - f|^p \right)^{\frac{q}{p}} \operatorname{meas}(\{|f_n - f| > k\})^{1-\frac{q}{p}} + \int_{\{|f_n-f|\le k\}} |f_n - f|^q .$$

Inequalities (3.2.1) and (3.2.2) imply then

$$\int_\Omega |f_n - f|^q \le L^q \left(\frac{L}{k}\right)^{p-q} + \int_{\{|f_n-f|\le k\}} |f_n - f|^q .$$

For every fixed $k \in \mathbb{R}^+$ Lebesgue's theorem implies that the second term of the right-hand side goes to 0, as $n \to \infty$. Moreover, for every fixed $\varepsilon > 0$ there exists k_ε such that the first term is smaller than ε, for $k \ge k_\varepsilon$. For such k_ε, there exists n_ε such that the second term is smaller than ε, for $n \ge n_\varepsilon$. In conclusion $f_n \to f$ in $L^q(\Omega)$ if $q < p$.

Let us prove that $f_n \to f$ weakly in $L^p(\Omega)$, $p \ge 1$. Since f_n is bounded in $L^p(\Omega)$, we can extract a subsequence which converges weakly in $L^p(\Omega)$. The limit is necessarily f, since $f_n \to f$ in $L^q(\Omega)$, $q < p$. To prove that $f_n \to f$ weakly in $L^p(\Omega)$ one can argue by contradiction. □

The following theorem gives us some sufficient conditions for the convergence in $L^p(\Omega)$.

Theorem 3.2 (Vitali). *Let f_n be a sequence of functions and f be a function in $L^p(\Omega)$. Assume that*

(1) *$f_n \to f$ a.e. in Ω;*
(2) *$\lim_{meas(E)\to 0} \int_E |f_n|^p = 0$, uniformly with respect to n, if E is a measurable subset of Ω.*

Then $f_n \to f$ in $L^p(\Omega)$.

Proof. Let us fix $\varepsilon > 0$. Let $E \subset \Omega$ be a measurable set; we have

$$\int_\Omega |f_n - f|^p \le \int_{\Omega\setminus E} |f_n - f|^p + 2^{p-1} \int_E [|f_n|^p + |f|^p] . \tag{3.2.3}$$

Using assumption (2), there exists $\delta_1(\varepsilon) > 0$ such that, if $\operatorname{meas}(E) < \delta_1(\varepsilon)$, then

$$\int_E |f_n|^p < \varepsilon , \quad \forall\, n \in \mathbb{N} .$$

Since $f \in L^p(\Omega)$ there exists $\delta_2(\varepsilon) > 0$ such that, if $\text{meas}(E) < \delta_2(\varepsilon)$, then

$$\int_E |f|^p < \varepsilon .$$

In conclusion the second term of the right-hand side of (3.2.3) is less than $2^p \varepsilon$. Let us study the first one. Setting $\delta = \min\{\delta_1(\varepsilon), \delta_2(\varepsilon)\}$ and using Egorov's theorem (Theorem 3.19), there exist $\nu_\varepsilon \in \mathbb{N}$ and a measurable set $E_0 \subset \Omega$ such that $\text{meas}(E_0) < \delta$ and

$$\int_{\Omega \setminus E_0} |f_n - f|^p < \varepsilon ,$$

for every $n > \nu_\varepsilon$. Choosing $E = E_0$ in (3.2.3), we get the result. □

The following result is a corollary of Vitali's theorem.

Theorem 3.3. *Let f_n be a sequence of functions and f be a function in $L^p(\Omega)$, $p \geq 1$. Then $f_n \to f$ in $L^p(\Omega)$ if and only if*
(1) *$f_n \to f$ in measure;*
(2) *$\lim_{meas(E) \to 0} \int_E |f_n|^p = 0$ uniformly with respect to n, where E is a measurable subset of Ω.*

Proof. We divide the proof into two parts.

Part I: assume that $f_n \to f$ in $L^p(\Omega)$. Clearly $f_n \to f$ in measure. Moreover, if E is any measurable subset of Ω, one has

$$\int_E |f_n|^p = \int_E |f_n - f + f|^p \leq 2^{p-1} \int_E |f_n - f|^p + 2^{p-1} \int_E |f|^p .$$

Let us fix $\varepsilon > 0$; since $f \in L^p(\Omega)$ there exists $\delta(\varepsilon) > 0$ such that, if $\text{meas}(E) < \delta(\varepsilon)$, then

$$\left[\int_E |f|^p\right]^{\frac{1}{p}} < \varepsilon . \tag{3.2.4}$$

On the other hand, since $f_n \to f$ in $L^p(\Omega)$, there exists ν_ε in $\mathbb{N}$ such that

$$\left(\int_E |f_n|^p\right)^{\frac{1}{p}} - \left(\int_E |f|^p\right)^{\frac{1}{p}} \leq \left(\int_E |f_n - f|^p\right)^{\frac{1}{p}} < \varepsilon, \quad \forall\, n > \nu_\varepsilon . \tag{3.2.5}$$

This implies that $\forall\, n > \nu_\varepsilon$, if $\text{meas}(E) < \delta(\varepsilon)$ then

$$\left(\int_E |f_n|^p\right)^{\frac{1}{p}} \leq \varepsilon + \left(\int_E |f|^p\right)^{\frac{1}{p}} \leq 2\varepsilon$$

by (3.2.4). Since $f_1, \dots, f_{\nu_\varepsilon} \in L^p(\Omega)$, there exists $\delta_1(\varepsilon)$ such that if $\text{meas}(E) < \delta_1(\varepsilon)$ then

$$\int_E |f_j|^p < \varepsilon, \quad \forall\, j = 1, \dots, \nu_\varepsilon .$$

This proves the result.

Part II: let us prove that assumptions (1) and (2) imply that $f_n \to f$ in $L^p(\Omega)$. Since $f_n \to f$ in measure, we can extract a subsequence such that $f_{n_k} \to f$ a.e. in Ω. Vitali's theorem (Theorem 3.2) implies that $f_{n_k} \to f$ in $L^p(\Omega)$. Indeed, the whole sequence f_n converges to f in $L^p(\Omega)$. If there exist a subsequence f_{n_j} and $\varepsilon_0 > 0$ such that

$$\|f_{n_j} - f\|_{L^p(\Omega)} \geq \varepsilon_0 , \tag{3.2.6}$$

using the previous argument, we could extract from f_{n_j} a subsequence converging to f in $L^p(\Omega)$. This is in contradiction with (3.2.6). □

We now give the definition of a Carathéodory function.

Definition 3.4. A function $g = g(x, \xi) : \Omega \times \mathbb{R}^m \to \mathbb{R}$ is a Carathéodory function if it is continuous with respect to ξ, for almost every x in Ω and measurable with respect to x for every ξ in $\mathbb{R}^m$.

Lemma 3.5. *Let $f(x,t) : \Omega \times \mathbb{R} \to \mathbb{R}$ be a Carathéodory function. Let u_n be a sequence of functions and u_0 be a measurable function such that $u_n \to u_0$ in measure. Then $f(x, u_n) \to f(x, u_0)$ in measure.*

Proof. Let $\varepsilon > 0$ and let u be any measurable function. We set, for $k > 0$

$$\Omega_k = \left\{ x \in \Omega : |u_0(x) - u(x)| < \frac{1}{k} \Rightarrow |f(x, u_0(x)) - f(x, u(x))| < \varepsilon \right\} .$$

Since f is continuous with respect to s, one has $\bigcup_{k \in \mathbb{N}} \Omega_k = \Omega$. Moreover $\lim_{k \to \infty} \text{meas}(\Omega_k) = \text{meas}(\Omega)$, as $\Omega_i \subset \Omega_j$, for $i < j$. Therefore, for any fixed $\eta > 0$, there exists k_0 such that

$$\text{meas}(\Omega) - \text{meas}(\Omega_{k_0}) < \frac{\eta}{2} .$$

Let us set

$$A_n = \left\{ x \in \Omega : |u_n(x) - u_0(x)| < \frac{1}{k_0} \right\} ;$$

since $u_n \to u_0$ in measure, there exists n_0 such that

$$\text{meas}(\Omega) - \text{meas}(A_n) < \frac{\eta}{2}$$

for every $n > n_0$. We set

$$D_n = \{ x \in \Omega : |f(x, u_n(x)) - f(x, u_0(x))| < \varepsilon \} .$$

By definition, one has $A_n \cap \Omega_{k_0} \subset D_n$ and this yields

$$\begin{aligned}\text{meas}(\Omega) - \text{meas}(D_n) &< [\text{meas}(\Omega) - \text{meas}(A_n)] + [\text{meas}(\Omega) - \text{meas}(\Omega_{k_0})] \\ &< \frac{\eta}{2} + \frac{\eta}{2} = \eta .\end{aligned}$$

This proves the result. □

We can now prove Nemitski's composition theorem.

Theorem 3.6 (Nemitski's composition theorem). *Let $p, q \geq 1$. Let $f(x,t) : \Omega \times \mathbb{R} \to \mathbb{R}$ be a Carathéodory function. Assume that there exist a positive function $a \in L^q(\Omega)$ and a constant $b > 0$ such that*

$$|f(x,t)| \leq a(x) + b|t|^{\frac{p}{q}} . \tag{3.2.7}$$

Then the operator

$$\begin{aligned}\Phi : \quad L^p(\Omega) &\mapsto L^q(\Omega) \\ u(x) &\mapsto f(x, u(x))\end{aligned}$$

is continuous.

Proof. Assume that $u_n \to u$ in $L^p(\Omega)$: we have to prove that $\Phi(u_n) \to \Phi(u)$ in $L^q(\Omega)$. We will prove that $\Phi(u_n)$ satisfies hypotheses 1 and 2 of Theorem 3.3. Clearly $u_n \to u$ in measure and so, using Lemma 3.5, $f(x, u_n(x)) \to f(x, u(x))$ in measure, that is, $\Phi(u_n) \to \Phi(u)$ in measure. Let us prove that

$$\lim_{\text{meas}(E) \to 0} \int_E |\Phi(u_n)|^q = 0$$

uniformly with respect to n. If E is a measurable subset of Ω, using (3.2.7), we get

$$\int_E |f(x, u_n(x))|^q \leq 2^{q-1} \int_E a(x)^q + 2^{q-1} b \int_E |u_n(x)|^p ,$$

and so $\lim_{\text{meas}(E)\to 0} \int_E |f(x, u_n(x))|^q = 0$, uniformly with respect to n. Indeed the first term tends to 0, since $a^q \in L^1(\Omega)$; the second term tends to 0 by Theorem 3.3 applied to the sequence u_n. Theorem 3.3 applied to the sequence $f(x, u_n(x))$ concludes the proof. □

We end this section with a result that we will use in Chapter 5.

Theorem 3.7. *Let $q > 1$ and $p \geq 1$. Let $f(x,t) : \Omega \times \mathbb{R} \to \mathbb{R}$ satisfy the same hypotheses as Theorem 3.6. If $u_n \to u$ weakly in $L^p(\Omega)$ and a.e. in Ω, then $f(x, u_n) \to f(x, u)$ weakly in $L^q(\Omega)$.*

Proof. It is clear that $f(x, u_n) \to f(x, u)$ a.e. in Ω. Since $\|u_n\|_{L^p(\Omega)}$ is bounded, due to (3.2.7), $\|f(x, u_n)\|_{L^q(\Omega)}$ is bounded too. Therefore, applying Theorem 3.1 to $f(x, u_n)$, we get that $f(x, u_n) \to f(x, u)$ weakly in $L^q(\Omega)$. □

3.3 Marcinkiewicz spaces

In this section, we will define a functional space that will be natural in the study of the regularity of the solutions to some differential problems.

Definition 3.8. Let Ω be a bounded open subset of $\mathbb{R}^N$. Let $p \geq 0$. The Marcinkiewicz space $M^p(\Omega)$ is the space of all measurable functions $f : \Omega \to \mathbb{R}$ with the following property: there exists a constant $\gamma > 0$ such that

$$\text{meas}(\{|f| > \lambda\}) \leq \frac{\gamma}{\lambda^p}, \quad \forall \lambda > 0. \tag{3.3.1}$$

The norm of $f \in M^p(\Omega)$ is defined by

$$\|f\|^p_{M^p(\Omega)} = \inf\{\gamma > 0 : \text{(3.3.1) holds}\} .$$

It is easy to see that $M^p(\Omega) \subseteq L^p(\Omega)$ for $p \geq 1$, as the following proposition shows:

Proposition 3.9. *Let $p \geq 1$. Then $L^p(\Omega) \subset M^p(\Omega)$.*

Proof. Let f be an $L^p(\Omega)$ function. Then

$$\int_\Omega |f|^p \geq \int_{\{|f|>\lambda\}} \lambda^p = \lambda^p \text{meas}(\{|f| > \lambda\}) ,$$

that is, f belongs to $M^p(\Omega)$. □

Remark 3.10. The inclusion $L^p(\Omega) \subseteq M^p(\Omega)$ is strict; indeed it is sufficient to consider the function $f(x) = \frac{1}{x}$ in $\Omega = (0,1) \subset \mathbb{R}$. This function does not belong to $L^1((0,1))$, but to $M^1((0,1))$, because

$$\text{meas}\left(\left\{\left|\frac{1}{x}\right| > \lambda\right\}\right) \leq \frac{1}{\lambda}.$$

We now want to prove that $M^p(\Omega)$ is included in some Lebesgue space. We will denote by A_k the set $\{|f| \geq k\}$, and by B_k the set $\{k \leq |f| < k+1\}$. The following lemma will prove to be useful:

Lemma 3.11. *Let $r \geq 1$ and $f : \Omega \to \mathbb{R}$ be a measurable function. $f \in L^r(\Omega)$ if and only if $\sum_{k=0}^\infty k^{r-1}$ meas$(A_k) < +\infty$.*

Proof. We begin with some simple remarks. Note that

$$\int_\Omega |f|^r = \sum_{k=0}^\infty \int_{B_k} |f|^r . \tag{3.3.2}$$

Moreover $A_k = \bigcup_{i=k}^\infty B_i$, and this union is disjoint; therefore

$$\sum_{k=0}^\infty k^{r-1}\text{meas}(A_k) = \sum_{k=0}^\infty k^{r-1} \sum_{i=k}^\infty \text{meas}(B_i) = \sum_{i=0}^\infty \text{meas}(B_i) \sum_{k=0}^i k^{r-1} . \tag{3.3.3}$$

Moreover, if $g : \mathbb{R}^+ \to \mathbb{R}^+$ is an increasing and continuous function, then

$$\sum_{k=0}^{n} g(k) \le \int_0^{n+1} g(t)\,dt \le \sum_{k=0}^{n} g(k+1)\,.$$

In particular, we may set $g(t) = t^{r-1}$, for $r \ge 1$; hence,

$$\sum_{j=0}^{m-1} j^{r-1} \le \int_0^m t^{r-1}\,dt = \frac{m^r}{r} \le \sum_{j=0}^{m-1} (j+1)^{r-1}\,. \tag{3.3.4}$$

Step I: Assume that $f \in L^r(\Omega)$. $\sum_{k=0}^{\infty} k^{r-1}\text{meas}(A_k) < +\infty$. Using the first inequality of (3.3.4) in (3.3.3), we get

$$\sum_{k=0}^{\infty} k^{r-1}\text{meas}(A_k) \le \sum_{i=0}^{\infty} \text{meas}(B_i)\,\frac{(i+1)^r}{r}\,.$$

Since $|f| \ge i$ over B_i, one has

$$\begin{aligned}\sum_{k=0}^{\infty} k^{r-1}\,\text{meas}(A_k) &\le \sum_{i=0}^{\infty} \frac{1}{r}\int_{B_i} (1+|f|)^r = \frac{1}{r}\int_{\Omega} (1+|f|)^r \\ &\le \frac{2^{r-1}}{r}\left[\text{meas}(\Omega) + \int_{\Omega} |f|^r\right]\end{aligned}$$

where we have used (3.3.2).

Step II: Assume that

$$\sum_{k=0}^{\infty} k^{r-1}\,\text{meas}(A_k) < +\infty\,.$$

We are going to prove that f belongs to $L^r(\Omega)$. One has, using (3.3.3) and the last inequality of (3.3.4)

$$\begin{aligned}\sum_{k=0}^{\infty} k^{r-1}\,\text{meas}(A_k) &= \sum_{i=0}^{\infty} \text{meas}(B_i) \sum_{k=0}^{i} k^{r-1} \\ &= \sum_{i=0}^{\infty} \text{meas}(B_i) \sum_{h=0}^{i-1} (h+1)^{r-1} \ge \sum_{i=0}^{\infty} \text{meas}(B_i)\,\frac{i^r}{r}\end{aligned}$$

and so

$$\sum_{i=0}^{\infty} \text{meas}(B_i)\, i^r < \infty\,.$$

By the definition of B_i

$$\sum_{i=0}^{\infty} \text{meas}(B_i)\, i^r \ge \sum_{i=2}^{\infty} \int_{B_i} (|f|-1)^r \ge \frac{1}{2^{r-1}} \int_{\Omega\setminus(B_0\cup B_1)} |f|^r - \text{meas}(\Omega)\,.$$

This implies that $f \in L^r(\Omega)$. □

Proposition 3.12. *Let $p > 1$ and $0 < \varepsilon \le p - 1$. Then $M^p(\Omega) \subset L^{p-\varepsilon}(\Omega)$.*

Proof. Let $f \in M^p(\Omega)$. Using Lemma 3.11, it suffices to prove that

$$\sum_{k=0}^{\infty} k^{p-\varepsilon-1}\,\mathrm{meas}(A_k) < \infty\,.$$

Since $\mathrm{meas}(A_k) \le \frac{\gamma}{k^p}$ for some $\gamma > 0$, one has

$$\sum_{k=0}^{\infty} k^{p-\varepsilon-1}\,\mathrm{meas}(A_k) \le \sum_{k=0}^{\infty} k^{p-\varepsilon-1}\,\frac{\gamma}{k^p}\,;$$

the last series is finite since $\varepsilon > 0$. This proves the result. □

Proposition 3.13. *Let f be an $M^p(\Omega)$ function, $p > 1$. Then there exists $B > 0$ such that, for every measurable set $E \subset \Omega$*

$$\int_E |f| \le B\,meas(E)^{1-\frac{1}{p}}\,, \tag{3.3.5}$$

where $B = B(\|f\|_{M^p(\Omega)}, p)$.

Proof. Once we prove that for every $f \in L^1(\Omega)$

$$\int_E |f| = \int_0^{+\infty} \mathrm{meas}(A_t \cap E)\,dt\,, \tag{3.3.6}$$

where E is any measurable set of Ω, the statement of the proposition is easy to prove. Indeed, if (3.3.6) holds, one has

$$\begin{aligned}
\int_E |f| &= \int_0^{+\infty} \mathrm{meas}(A_t \cap E)\,dt = \\
&= \int_0^{\mathrm{meas}(E)^{-\frac{1}{p}}} \mathrm{meas}(A_t \cap E)\,dt + \int_{\mathrm{meas}(E)^{-\frac{1}{p}}}^{+\infty} \mathrm{meas}(A_t \cap E)\,dt \\
&\le \mathrm{meas}(E)^{1-\frac{1}{p}} + \int_{\mathrm{meas}(E)^{-\frac{1}{p}}}^{+\infty} \mathrm{meas}(A_t)\,dt \\
&\le \mathrm{meas}(E)^{1-\frac{1}{p}} + \|f\|^p_{M^p(\Omega)} \int_{\mathrm{meas}(E)^{-\frac{1}{p}}}^{+\infty} t^{-p}\,dt \\
&\le B\,\mathrm{meas}(E)^{1-\frac{1}{p}}\,,
\end{aligned}$$

where B depends on $\|f\|_{M^p(\Omega)}$ and on p.

Step I: Assume that $f(x) = \alpha\chi_E(x), \alpha > 0$. Then $\int_E |f(x)| = \alpha\,\mathrm{meas}(E)$. On the other hand,

$$A_t \cap E = \{x \in E : |f(x)| > t\} = \begin{cases} E, & \text{if } t \le \alpha, \\ \emptyset, & \text{if } t > \alpha, \end{cases} \tag{3.3.7}$$

and so

$$\int_0^{+\infty} \mathrm{meas}(A_t \cap E)\,dt = \int_0^{\alpha} \mathrm{meas}(E)\,dt = \alpha\,\mathrm{meas}(E)\ .$$

Therefore (3.3.6) holds true for $f(x) = \alpha\chi_E(x), \alpha > 0$.

Step II: Assume that $f(x) = \sum_{i=1}^{M} \alpha_i\chi_{E_i}$, where E_i are measurable subsets of E such that

$$\bigcup_{i=1}^{M} E_i = E\,, \quad E_i \cap E_j = \emptyset \text{ se } i \neq j\,,$$

$\alpha_i \in \mathbb{R}^+$ and $M \in \mathbb{N}$. From Step I, one has

$$\int_E |f| = \sum_{i=1}^{M} \int_E \alpha_i\chi_{E_i} = \sum_{i=1}^{M} \int_0^{+\infty} \mathrm{meas}(A_{i,t})\,dt$$

where $A_{i,t} = \{x \in \Omega : \alpha_i\chi_{E_i}(x) > t\}$ for $i \in \{1, \dots, M\}$. Now,

$$A_t \cap E = \bigcup_{i=1}^{M}(A_t \cap E_i) = \bigcup_{i=1}^{M}\left\{x \in E_i : \sum_{i=1}^{M} \alpha_i\chi_{E_i} > t\right\} = \bigcup_{i=1}^{M} A_{i,t}\ .$$

This implies that

$$\sum_{i=1}^{M} \int_0^{+\infty} \mathrm{meas}(A_{i,t})\,dt = \int_0^{+\infty} \sum_{i=1}^{M} \mathrm{meas}(A_{i,t})\,dt = \int_0^{+\infty} \mathrm{meas}(A_t \cap E)\,dt\ .$$

Therefore, (3.3.6) holds true for positive step functions.

Step III: Let f be any function in $L^1(\Omega)$. There exists a sequence of positive step functions such that

$$s_n(x) \nearrow |f(x)| \quad \text{a.e. in } E\,.$$

From Beppo Levi's theorem and from Step II, one has

$$\begin{aligned}\int_E |f| &= \lim_{n\to\infty} \int_E s_n = \lim_{n\to\infty} \int_0^{+\infty} \mathrm{meas}\{x \in E : |s_n(x)| > t\}\,dt \\ &= \lim_{n\to\infty} \int_0^{+\infty} dt \int_E \chi_{\{x\in E:|s_n(x)|>t\}}\,.\end{aligned} \tag{3.3.8}$$

Lebesgue's theorem implies that

$$\lim_{n\to\infty}\int_E \chi_{\{x\in E:|s_n(x)|>t\}} = \int_E \chi_{A_t\cap E}\,. \tag{3.3.9}$$

We set

$$g_n(t) = \int_E \chi_{\{x\in E:|s_n(x)|>t\}}\,.$$

Then (3.3.8) is equivalent to

$$\int_E |f| = \lim_{n\to\infty}\int_0^{+\infty} g_n(t)\,dt$$

and (3.3.9) means that $g_n(t) \to \text{meas}(A_t\cap E)$ if $n\to\infty$. To prove the result it suffices to prove that

$$\lim_{n\to+\infty}\int_0^{+\infty} g_n(t)\,dt = \int_0^{+\infty}\text{meas}(A_t\cap E)\,dt\,.$$

We have that $g_n(t) \to \text{meas}(A_t\cap E)$ a.e. in $(0,+\infty)$; moreover $|g_n(t)| \le \text{meas}(E)$; it is sufficient to prove that $\text{meas}(A_t\cap E)$ belongs to $L^1((0,+\infty))$, and then apply Lebesgue's theorem. This is easy, since

$$\begin{aligned}\int_0^{+\infty}\text{meas}(A_t\cap E)dt &\le \int_0^{1}\text{meas}(A_t\cap E)\,dt + \int_1^{+\infty}\text{meas}(A_t)\,dt\\ &\le \text{meas}(E) + \int_1^{+\infty}\frac{\|f\|^p_{M^p(\Omega)}}{t^p}\,dt < +\infty\,.\end{aligned}$$

Therefore (3.3.6) is proved. □

3.4 Appendix

We recall here the results on Lebesgue spaces that we use in this book (see [29] for more details).

Let E be a Lebesgue measurable subset of $\mathbb{R}^N, N \ge 1$. Let $1 < p < \infty$; p' will denote the number $\frac{p}{p-1}$.

Theorem 3.14 (Hölder's inequality). *Let $f \in L^p(E)$ and $g \in L^{p'}(E)$. Then*

$$\int_E fg \le \|f\|_{L^p(E)}\|g\|_{L^{p'}(E)}\,.$$

Theorem 3.15 (Interpolation inequality). *Let $p, q, r \in [1, +\infty)$ such that $p < r < q$. Let $f \in L^q(E)$. Then*

$$\|f\|_{L^r(E)} \leq \|f\|_{L^p(E)}^{\theta} \|f\|_{L^q(E)}^{1-\theta} ,$$

where θ is such that $\frac{1}{r} = \frac{\theta}{p} + \frac{1-\theta}{q}$.

Theorem 3.16 (Beppo Levi). *Let f_n be a sequence of $L^1(E)$ functions such that*
(1) *$0 \leq f_n(x) \leq f_{n+1}(x)$ a.e. in E for every $n \in \mathbb{N}$;*
(2) *$\int_E f_n < +\infty$ for every $n \in \mathbb{N}$.*

Then $f_n \to f$ in $L^1(E)$.

Theorem 3.17 (Lebesgue). *Let f_n be a sequence of $L^1(E)$ functions such that*
(1) *$f_n \to f$ a.e. in E;*
(2) *there exists $g \in L^1(E)$ such that $|f_n(x)| \leq g(x)$ a.e. in E.*

Then $f_n \to f$ in $L^1(E)$.

Theorem 3.18 (Fatou). *Let f_n be a sequence of $L^1(E)$ functions such that*
(1) *$f_n \geq 0$ a.e. in E;*
(2) *$\int_E f_n < +\infty$ for every $n \in \mathbb{N}$.*

Let $f(x) = \liminf_{n\to\infty} f_n(x)$ for a.e. $x \in E$. Then $\int_E f \leq \liminf_{n\to\infty} \int_E f_n$.

Theorem 3.19 (Egorov). *Let f_n be a sequence of functions and f be a function defined on E, with meas$(E) < +\infty$. Assume that $f_n \to f$ a.e. in E. Then for every $\varepsilon > 0$ there exists a measurable subset A of E such that meas$(E \setminus A) < \varepsilon$ and $f_n \to f$ uniformly on A, as $n \to \infty$.*

Theorem 3.20. *Let $1 < p < \infty$. A sequence f_n of $L^p(E)$ functions converges weakly to f in $L^p(E)$ if $\int_E (f_n - f)\, g \to 0$ for every $g \in L^{p'}(E)$. A sequence f_n of $L^1(E)$ functions converges weakly to f in $L^1(E)$ if $\int_E (f_n - f) g \to 0$ for every $g \in L^\infty(E)$.*

Theorem 3.21. *Let $1 < p < \infty$. Every bounded sequence f_n in $L^p(E)$ has a subsequence weakly converging to some $f \in L^p(E)$.*

Theorem 3.22 (Dunford–Pettis). *Let f_n be a bounded sequence of $L^1(E)$ functions. Assume that for every measurable subset $A \subset E$, one has $\int_A |f_n| \to 0$, as meas$(A) \to 0$, uniformly with respect to n. Then f_n has a subsequence weakly converging to some $f \in L^1(E)$.*

Definition 3.23. We recall that f_n converges to f in measure in E if meas$(\{x \in E : |f_n - f| \geq \varepsilon\}) \to 0$ as $\varepsilon \to 0$.

4 Linear and semilinear elliptic equations

4.1 Introduction

In this chapter, we will study the existence of solutions to linear and semilinear elliptic problems. More precisely, let Ω be an open bounded subset of $\mathbb{R}^N$, $N > 3$. We assume that $M(x)$ is an $N \times N$ symmetric matrix with the following properties:
(1) M is elliptic, that is, there exists $\alpha > 0$ such that $M(x)\xi \cdot \xi \geq \alpha|\xi|^2$, $\forall\, \xi \in \mathbb{R}^N$;
(2) M is bounded, meaning there exists $\beta > 0$ such that $|M(x)| \leq \beta\,, \quad \forall\, x \in \Omega$.

The first class of problems that we will consider is linear:

$$\begin{cases} -\operatorname{div}(M(x)\nabla u) = f\,, & \text{in} \quad \Omega\,, \\ u = 0\,, & \text{on} \quad \partial\Omega\,. \end{cases}$$

Afterward we will add a nonlinearity in u of the form

$$\begin{cases} -\operatorname{div}(M(x)\nabla u) + g(u) = f\,, & \text{in} \quad \Omega\,, \\ u = 0\,, & \text{on} \quad \partial\Omega\,. \end{cases}$$

The existence results that we will present are based on Functional Analysis results by Lax, Milgram, and Stampacchia (see [32, 48, 51]).

4.2 The Lax–Milgram and Stampacchia's theorems

In this section, we present the Lax–Milgram theorem and Stampacchia's theorem. They will be useful to study the elliptic problems of this chapter.

We will use the following notations. Let H be a Hilbert space: $(u|v)$ will denote the scalar product of two elements $u, v \in H$, and $\|u\| = \sqrt{(u|u)}$ will denote the norm of $u \in H$. If $\varphi \in H'$, $\langle \varphi, v\rangle$ will be the value of φ at $v \in H$.

Proposition 4.1 (Stampacchia). *Let H be a Hilbert space and $K \subseteq H$ a closed convex set. Let $a : K \times K \to \mathbb{R}$ be a bilinear, continuous, coercive form, that is, there exist $\alpha, \beta > 0$ such that*

$$a(u, v) \leq \beta\|u\|\|v\|$$

and

$$a(u, u) \geq \alpha\|u\|\|v\|$$

for every $u, v \in H$. Then, for every fixed $g \in H'$ there exists only one $u \in K$ such that

$$a(u, v - u) \geq \langle g, v - u\rangle\,, \quad \forall\, v \in K\,.$$

Proof. For $g \in H'$ fixed, Riesz theorem (Theorem 4.16) posits the existence of a unique $f \in H$ such that

$$\langle g, v\rangle = (f|v)\,, \quad \forall\, v \in H\,.$$

Moreover, for any fixed u, the map $v \mapsto a(u,v)$ is continuous and linear on H. Theorem 4.16 again implies that there exists $A(u) \in H$ such that $a(u,v) = (A(u)|v)$. The proposition will follow once we prove that there exists a unique $u \in K$ such that

$$(A(u) - f|v-u) \ge 0\,, \quad \forall\, v \in K\,,$$

that is,

$$(-\lambda\, A(u) + \lambda\, f|v-u) \le 0\,, \quad \forall\, v \in K\,,$$

for some $\lambda > 0$. Let $C : H \to K \subset H$ be the map which associates with $z \in H$ the projection over K of $z - \lambda A(z) + \lambda f$. Property (4.6.1) of the projection implies that

$$(z - \lambda A(z) + \lambda f - C(z)|v - C(z)) \le 0\,, \quad \forall\, v \in K\,.$$

To prove the result, it is sufficient to find a fixed point of C. Using property (4.6.2) of the projection one has

$$\|C(z_1) - C(z_2)\|^2 \le \|z_1 - z_2 - \lambda(A(z_1) - A(z_2))\|^2\,.$$

Since A is continuous and coercive, the following estimates hold:

$$\begin{aligned}\|C(z_1) - C(z_2)\|^2 &\le \|z_1 - z_2\|^2 + \lambda^2\|A(z_1) - A(z_2)\|^2 \\ &\quad - 2\lambda\,(z_1 - z_2|A(z_1) - A(z_2)) \\ &\le \|z_1 - z_2\|^2 + \lambda^2\beta^2\|z_1 - z_2\|^2 - 2\alpha\lambda\,\|z_1 - z_2\|^2 \\ &= (1 + \lambda^2\beta^2 - 2\alpha\lambda)\|z_1 - z_2\|^2\,.\end{aligned}$$

Therefore C is a contraction if $0 < \lambda < 2\alpha/\beta^2$. By Theorem 2.1 C has a unique fixed point. □

Theorem 4.2 (Lax–Milgram). *Let H be an Hilbert space and $a : H \times H \to \mathbb{R}$ a bilinear, continuous, coercive form. Then, for every fixed $\varphi \in H'$, there exists a unique $u \in H$ such that*

$$a(u,v) = \langle \varphi, v\rangle\,, \quad \forall\, v \in H\,.$$

Proof. Let $\varphi \in H'$; by Proposition 4.1 there exists a unique $u \in H$ such that

$$a(u, v-u) \ge \langle \varphi, v-u\rangle\,, \quad \forall\, v \in H\,.$$

In particular

$$a(u, t\,v-u) \ge \langle \varphi, t\,v-u\rangle\,, \quad \forall\, v \in H\,, \quad \forall\, t \in \mathbb{R}\,,$$

so that

$$t\,[a(u,v) - \langle\varphi, v\rangle] \geq a(u,u) - \langle\varphi, u\rangle\,, \qquad \forall\, t \in \mathbb{R}\,.$$

Hence

$$a(u,v) - \langle\varphi, v\rangle = 0$$

for every $v \in H$. □

Theorem 4.3 (Stampacchia). *Let H be a Hilbert space. Let $a : H \times H \to \mathbb{R}$ be a continuous and linear form in the second variable such that*

(1) $|a(\psi_1, w) - a(\psi_2, w)| \leq \beta\, \|\psi_1 - \psi_2\|\, \|w\|\,, \quad \forall\, \psi_1, \psi_2, w \in H;$

(2) $a(\psi_1, \psi_1 - \psi_2) - a(\psi_2, \psi_1 - \psi_2) \geq C\, \|\psi_1 - \psi_2\|^2\,, \quad \forall\, \psi_1, \psi_2 \in H.$

Then, for every $\varphi \in H'$ there exists a unique $u \in H$ such that $a(u,w) = \varphi(w)$ for every $w \in H$.

Proof. We divide the proof into two steps.

Step I: Let us prove that if $A : H \to H$ satisfies for some positive α, γ

(1) $\|A(x) - A(y)\| \leq \gamma\, \|x - y\|\,, \quad \forall\, x, y \in H$

(2) $(x - y | A(x) - A(y)) \geq \alpha\, \|x - y\|^2\,, \quad \forall\, x, y \in H$

then, for every fixed $f \in H$, there exists a unique $u \in H$ such that $A(u) = f$. For this it is sufficient to prove that

$$R(v) = v - \lambda\, A(v) + \lambda\, f$$

is a contraction for some λ. By definition

$$\begin{aligned}\|R(v) - R(w)\|^2 &= (R(v) - R(w) | R(v) - R(w)) \\ &= \|v - w\|^2 + \lambda^2 \|A(v) - A(w)\|^2 - 2\lambda\, (v - w | A(v) - A(w))\,.\end{aligned}$$

By the hypotheses on A, one has

$$\|R(v) - R(w)\|^2 \leq (1 + \lambda^2\gamma^2 - 2\lambda\,\alpha)\, \|v - w\|^2\,.$$

Thus R is a contraction as soon as $0 < \lambda < \frac{2\alpha}{\gamma^2}$.

Step II: By Riesz theorem (Theorem 4.16), for every fixed $\varphi \in H'$ there exists a unique $f_0 \in H$ such that

$$\varphi(w) = (f_0 | w)\,, \qquad \forall\, w \in H\,.$$

To prove the theorem we must find $u \in H$ such that

$$a(u,w) = (f_0 | w)$$

for every $w \in H$. Now, for every $v \in H$ we can define the following linear continuous functional on H:

$$\begin{aligned} T_v: \quad & H \to \mathbb{R} \\ & w \to a(v, w) . \end{aligned}$$

By Theorem 4.16 again, there exists a unique $v_0 \in H$ such that $T_v(w) = a(v, w) = (v_0|w)$ for every $w \in H$. The operator

$$\begin{aligned} A: \quad & H \to H \\ & v \to v_0 \end{aligned}$$

satisfies inequalities (1) and (2) of Step I. Consequently, given $f \in H$ there exists a unique u such that $A(u) = f$, that is, $a(u, w) = (A(u)|w) = (f|w)$ for every $w \in H$. Therefore there exists a unique u such that $a(u, w) = (f_0|w) = \varphi(w)$ for every $w \in H$. □

4.3 Linear equations

In this section we consider the linear problem

$$\begin{cases} -\mathrm{div}(M(x)\nabla u) = f, & \text{in } \Omega, \\ u = 0, & \text{on } \partial\Omega; \end{cases} \tag{4.3.1}$$

we remark that on taking M as the identity matrix, one has the Dirichlet problem for the Laplacian operator:

$$\begin{cases} -\Delta u = f, & \text{in } \Omega, \\ u = 0, & \text{on } \partial\Omega. \end{cases}$$

The proof of the existence and uniqueness of a solution to problem (4.3.1) is based on Lax–Milgram theorem.

Theorem 4.4. *Let $f \in L^m(\Omega)$, $m \geq \frac{2N}{N+2}$. Then there exists a unique weak solution $u \in H_0^1(\Omega)$ to problem (4.3.1). In other words,*

$$\int_\Omega M(x)\nabla u \cdot \nabla v = \int_\Omega f v, \quad \forall\, v \in H_0^1(\Omega).$$

Proof. Define $a : H_0^1(\Omega) \times H_0^1(\Omega) \to \mathbb{R}$ by

$$a(u, v) = \int_\Omega M(x)\nabla u \cdot \nabla v.$$

It is easily seen that a is continuous. Indeed, since M is bounded by hypothesis 2, the Cauchy–Schwarz inequality gives

$$|a(u, v)| \leq \int_\Omega |M(x)\nabla u \cdot \nabla v| \leq \beta\, \|\nabla u\|_{L^2(\Omega)}\, \|\nabla v\|_{L^2(\Omega)}, \quad \forall\, u, v \in H_0^1(\Omega).$$

On the other hand, a is coercive, since $M(x)\xi \cdot \xi \geq \alpha|\xi|^2$ and so

$$\int_\Omega M(x)\nabla u \cdot \nabla u \geq \alpha \int_\Omega |\nabla u|^2\,, \qquad \forall\, u \in H_0^1(\Omega)\,.$$

The result follows from the Lax–Milgram theorem. □

Remark 4.5. We will see a different proof of this theorem in Chapter 9.

4.4 Some semilinear monotone equations

In this section we study the following semilinear problem

$$\begin{cases} -\operatorname{div}(M(x)\nabla u) + g(u) = f\,, & \text{in} \quad \Omega\,, \\ u = 0\,, & \text{on} \quad \partial\Omega\,. \end{cases} \tag{4.4.1}$$

Theorem 4.6. *Let $g : \mathbb{R} \to \mathbb{R}$ be an increasing function. Suppose that g is Lipschitz continuous, that is, there exists a positive constant C such that*

$$|g(s) - g(t)| \leq C\,|s - t|\,, \quad \forall \quad s, t \in \mathbb{R}\,. \tag{4.4.2}$$

Let $f \in L^m(\Omega)$, $m \geq \frac{2N}{N+2}$. Then there exists a unique solution $u \in H_0^1(\Omega)$ to problem (4.4.1) in the following sense:

$$\int_\Omega M(x)\nabla u \cdot \nabla v + \int_\Omega g(u)v = \int_\Omega fv\,, \quad \forall\, v \in H_0^1(\Omega)\,.$$

Proof. Define the following form on $H_0^1(\Omega) \times H_0^1(\Omega)$:

$$a(u, w) = \int_\Omega M(x)\nabla u \cdot \nabla w + \int_\Omega g(u)w\,.$$

By the hypotheses on M and (4.4.2), one has

$$|a(u, w)| \leq \beta \int_\Omega |\nabla u|\,|\nabla w| + \int_\Omega [C\,|u| + g(0)]\,|w|\,,$$

that is, a is well defined. The form a is continuous and linear in the second variable. Indeed, if $w_n \to w$ in $H_0^1(\Omega)$, then

$$\int_\Omega M(x)\nabla u \cdot \nabla w_n \to \int_\Omega M(x)\nabla u \cdot \nabla w$$

and

$$\int_\Omega g(u)w_n \to \int_\Omega g(u)w$$

since $w_n \to w$ weakly in $L^2(\Omega)$ and $|g(u)| \le [C|u| + g(0)] \in L^2(\Omega)$. In other words, $a(u, w_n) \to a(u, w)$. Moreover, a satisfies hypothesis 1 of Theorem 4.3 because

$$\begin{aligned}|a(u_1, w) - a(u_2, w)| &= \left| \int_\Omega M(x)\nabla(u_1 - u_2) \cdot \nabla w + \int_\Omega [g(u_1) - g(u_2)]w \right| \\ &\le \beta \|\nabla(u_1 - u_2)\|_{L^2(\Omega)} \|\nabla w\|_{L^2(\Omega)} \\ &\quad + C\|u_1 - u_2\|_{L^2(\Omega)} \|w\|_{L^2(\Omega)}\end{aligned}$$

the last inequality following from hypothesis 2 on M and hypothesis (4.4.2) on g. Finally a satisfies hypothesis 2 of Theorem 4.3:

$$\begin{aligned}a(u_1, u_1 - u_2) - a(u_2, u_1 - u_2) &= \int_\Omega M(x)\nabla(u_1 - u_2) \cdot \nabla(u_1 - u_2) \\ &\quad + \int_\Omega [g(u_1) - g(u_2)](u_1 - u_2) \\ &\ge \alpha\|\nabla(u_1 - u_2)\|^2_{L^2(\Omega)}\end{aligned}$$

since M is elliptic and g is increasing. The result follows from Theorem 4.3. □

Theorem 4.7. *Let $g : \mathbb{R} \to \mathbb{R}$ be an increasing locally Lipschitz continuous function. Let $f \in L^m(\Omega)$, $m \ge \frac{2N}{N+2}$. Then there exists a unique solution $u \in H_0^1(\Omega)$ to problem (4.4.1) in the following sense:*

$$\int_\Omega M(x)\nabla u \cdot \nabla v + \int_\Omega g(u)v = \int_\Omega fv, \quad \forall\, v \in H_0^1(\Omega) \cap L^\infty(\Omega).$$

Moreover, $g(u) \in L^1(\Omega)$.

In the proof we make use of the following function defined for $k > 0$:

$$T_k(s) = \begin{cases} -k, & s \le -k, \\ s, & |s| \le k, \\ k, & s \ge k. \end{cases} \tag{4.4.3}$$

Proof. Step I: We prove the existence of a solution by approximation. To this end, let $g_n(t) = T_n(g(t))$; let $u_n \in H_0^1(\Omega)$ be the solutions to problems

$$\begin{cases} -\mathrm{div}(M(x)\nabla u_n) + g_n(u_n) = f, & \text{in } \Omega, \\ u_n = 0, & \text{on } \partial\Omega. \end{cases} \tag{4.4.4}$$

Such solutions exist due to Theorem 4.6, since g_n is increasing and Lipschitz continuous. Considering u_n as a test function in (4.4.4) one gets

$$\int_\Omega M(x)\nabla u_n \cdot \nabla u_n + \int_\Omega u_n g_n(u_n) = \int_\Omega f u_n.$$

Hölder's inequality on the right-hand side implies

$$\int_\Omega M(x)\nabla u_n \cdot \nabla u_n + \int_\Omega u_n g_n(u_n) \le \|f\|_{L^{\frac{2N}{N+2}}(\Omega)} \|u_n\|_{L^{2^*}(\Omega)} . \tag{4.4.5}$$

The ellipticity of M and the monotonicity of g used on the left-hand side of the above inequality give

$$\alpha\|\nabla u_n\|^2_{L^2(\Omega)} \le \|f\|_{L^{\frac{2N}{N+2}}(\Omega)} \|u_n\|_{L^{2^*}(\Omega)} .$$

By Sobolev's inequality on the left-hand side, $\|u_n\|_{H_0^1(\Omega)}$ is uniformly bounded. We deduce the existence of a $H_0^1(\Omega)$ function u such that $u_n \to u$ weakly in $H_0^1(\Omega)$ and a.e., up to a subsequence. Moreover, since $\int_\Omega M(x)\nabla u_n \cdot \nabla u_n \ge 0$, we deduce from (4.4.5) that there exists a constant $C > 0$ such that

$$\int_\Omega u_n g_n(u_n) \le C \tag{4.4.6}$$

for every n. Let us prove now that $g_n(u_n) \to g(u)$ in $L^1(\Omega)$. It is clear that $g_n(u_n) \to g(u)$ a.e. in Ω by the continuity of g. Moreover, if E is any subset of Ω and $t \in \mathbb{R}^+$ one has

$$\begin{aligned}\int_E |g_n(u_n)| &= \int_{\{x\in E:|u_n(x)|\le t\}} |g_n(u_n)| + \int_{\{x\in E:|u_n(x)|>t\}} |g_n(u_n)| \\ &\le \int_E |g_n(t)| + \frac{1}{t}\int_{\{x\in E:|u_n(x)|>t\}} u_n g(u_n) \\ &\le |g(t)|\,\text{meas}(E) + \frac{C}{t}\end{aligned}$$

due to (4.4.6). Consequently

$$\lim_{\text{meas}(E)\to 0} \int_E |g_n(u_n)| \le \frac{C}{t}, \quad \forall\, t > 0.$$

By Theorem 3.2, $g_n(u_n) \to g(u)$ in $L^1(\Omega)$. Therefore for every $\varphi \in H_0^1(\Omega) \cap L^\infty(\Omega)$ we can pass to the limit in

$$\int_\Omega M(x)\nabla u_n \cdot \nabla\varphi + \int_\Omega g_n(u_n)\varphi = \int_\Omega f\varphi$$

to get

$$\int_\Omega M(x)\nabla u \cdot \nabla\varphi + \int_\Omega g(u)\varphi = \int_\Omega f\varphi$$

for every $\varphi \in H_0^1(\Omega) \cap L^\infty(\Omega)$.

Step II: Let us prove that the solution to problem (4.4.1) is unique. By contradiction, if u_1 and u_2 are two solutions, then

$$\int_\Omega M(x)\nabla u_1 \cdot \nabla T_k(u_1 - u_2) + \int_\Omega g(u_1) T_k(u_1 - u_2) = \int_\Omega f\, T_k(u_1 - u_2);$$
$$\int_\Omega M(x)\nabla u_2 \cdot \nabla T_k(u_1 - u_2) + \int_\Omega g(u_2) T_k(u_1 - u_2) = \int_\Omega f\, T_k(u_1 - u_2).$$

This implies that

$$\int_\Omega M(x)\nabla(u_1 - u_2) \cdot \nabla T_k(u_1 - u_2) = \int_\Omega M(x)\nabla T_k(u_1 - u_2) \cdot \nabla T_k(u_1 - u_2) \le 0$$

by the monotonicity of g. Since M is elliptic, $T_k(u_1 - u_2) = 0$ a.e. for every k; therefore $u_1 = u_2$ a.e. in Ω. □

Example 4.8. Using the previous theorem one sees immediately that there exists a solution u to

$$u \in H_0^1(\Omega) \cap L^{p-1}(\Omega): \qquad -\mathrm{div}(M(x)\nabla u) + |u|^{p-2}u = fu \in H_0^1(\Omega),$$
$$e^u - 1 \in L^1(\Omega): \qquad -\mathrm{div}(M(x)\nabla u) + e^u - 1 = f.$$

Remark 4.9. In Chapters 10 and 11, we shall again study approximating problems to get a priori estimates and then pass to the limit.

4.5 Sub and supersolutions method

In this section we will study the semilinear problem

$$\begin{cases} -\mathrm{div}(M(x)\nabla u) = g(u) + f, & \text{in } \Omega, \\ u = 0, & \text{on } \partial\Omega, \end{cases} \tag{4.5.1}$$

under the following hypotheses:
(1) $f \in L^{\frac{2N}{N+2}}(\Omega)$;
(2) $g : \mathbb{R} \to \mathbb{R}$ is increasing and continuous; there exists $\gamma > 0$ such that

$$|g(s)| \le \gamma |s|^a, \quad a \le \frac{N+2}{N-2}.$$

We recall that
(1) M is elliptic, that is, there exists $\alpha > 0$ such that $M(x)\xi \cdot \xi \ge \alpha|\xi|^2$, $\forall\, \xi \in \mathbb{R}^N$;
(2) M is bounded, meaning there exists $\beta > 0$ such that $|M(x)| \le \beta$, $\forall\, x \in \Omega$.

We remark that as in the previous section g is assumed to be increasing, but here it appears on the right-hand side. We will solve this problem using the sub and super-solutions method (see [45]).

Definition 4.10. A function $\underline{u} \in H_0^1(\Omega)$ is a subsolution to problem (4.5.1) if for every positive $v \in H_0^1(\Omega)$

$$\int_\Omega M(x)\nabla\underline{u} \cdot \nabla v \le \int_\Omega g(\underline{u})v + \int_\Omega fv\,.$$

A function $\overline{u} \in H_0^1(\Omega)$ is a supersolution to problem (4.5.1) if for every positive $v \in H_0^1(\Omega)$

$$\int_\Omega M(x)\nabla\overline{u} \cdot \nabla v \ge \int_\Omega g(\overline{u})v + \int_\Omega fv\,.$$

Theorem 4.11. *Under the previous hypotheses, let $\underline{u}$ and $\overline{u}$ be a sub and a supersolution to (4.5.1) such that $\underline{u} \le \overline{u}$ a.e. in Ω. Then there exists $u \in H_0^1(\Omega)$ to problem (4.5.1); moreover $\underline{u} \le u \le \overline{u}$ a.e. in Ω.*

Theorem 4.12 (Maximum Principle). *Let $u \in H_0^1(\Omega)$.*
(1) *Assume that $\int_\Omega M(x)\nabla u \cdot \nabla v \le 0$ for every positive $v \in H_0^1(\Omega)$. Then $u \le 0$.*
(2) *Assume that $\int_\Omega M(x)\nabla u \cdot \nabla v \ge 0$ for every positive $v \in H_0^1(\Omega)$. Then $u \ge 0$.*

Proof. (1) Choose $v = u^+$, that is, the positive part of u as a test function. Then

$$\int_\Omega M(x)\nabla u^+ \cdot \nabla u^+ = \int_\Omega M(x)\nabla(u^+ - u^-) \cdot \nabla u^+ \le 0\,.$$

Using the ellipticity of M, $u^+ = 0$.
(2) Choose $v = u^-$, that is, the negative part of u as a test function. Then

$$0 \le \int_\Omega M(x)\nabla(u^+ - u^-) \cdot \nabla u^- = -\int_\Omega M(x)\nabla u^- \cdot \nabla u^-\,.$$

The ellipticity of M gives $u^- = 0$. □

We can now prove Theorem 4.11.

Proof. The proof is divided into two steps: in the first one we construct a sequence u_n in $H_0^1(\Omega)$ by induction and prove that $\underline{u} \le u_1 \le \cdots \le u_n \le \cdots \le \overline{u}$; in the second one we prove that this sequence converges to a solution u to problem (4.5.1).

Step I: Set $u_1 = \underline{u}$; let $u_n \in H_0^1(\Omega)$ be the solution to

$$\begin{cases} -\mathrm{div}(M(x)\nabla u_n) = g(u_{n-1}) + f\,, & \text{in } \Omega\,,\\ u_n = 0\,, & \text{on } \partial\Omega\,. \end{cases}$$

Such a solution exists by Theorem 4.4, since $g(u_{n-1}) + f \in L^{\frac{2N}{N+2}}(\Omega)$. Let us prove by induction that the sequence u_n is increasing. For every $\varphi \geq 0$, by the hypotheses and the definition of u_2, one has

$$\int_\Omega M(x)\nabla u_1 \cdot \nabla\varphi \leq \int_\Omega [g(u_1) + f]\,\varphi$$

$$\int_\Omega M(x)\nabla u_2 \cdot \nabla\varphi = \int_\Omega [g(u_1) + f]\,\varphi\,.$$

Consequently

$$\int_\Omega M(x)\nabla(u_1 - u_2) \cdot \nabla\varphi \leq 0\,.$$

Theorem 4.12 implies that $u_1 - u_2 \leq 0$.

Let us prove that $u_n \leq u_{n+1}$, assuming that $u_n \geq u_{n-1}$. We have

$$\int_\Omega M(x)\nabla u_n \cdot \nabla\varphi = \int_\Omega [g(u_{n-1}) + f]\,\varphi$$

$$\int_\Omega M(x)\nabla u_{n+1} \cdot \nabla\varphi = \int_\Omega [g(u_n) + f]\,\varphi\,.$$

Consequently

$$\int_\Omega M(x)\nabla(u_n - u_{n+1}) \cdot \nabla\varphi = \int_\Omega [g(u_{n-1}) - g(u_n)]\,\varphi.$$

Since g is increasing, $g(u_{n-1}) \leq g(u_n)$ and so $u_n \leq u_{n+1}$ by Theorem 4.12. Therefore u_n is increasing. With the same technique one can prove that $u_n \leq \overline{u}$.

Step II: We want to prove that u_n converges to a solution to problem (4.5.1). Let us first prove that the u_n converges to some $u \in L^{2^*}(\Omega)$. Since u_n is increasing and $u_n \leq \overline{u}$, $u_n(x)$ has a limit a.e., say $u(x)$. It follows from Step *I* that $|u_n| \leq |\underline{u}| + |\overline{u}|$; passing to the limit, one has $|u| \leq |\underline{u}| + |\overline{u}| \in L^{2^*}(\Omega)$. Hence, for a positive constant $C(N)$

$$|u_n - u|^{2^*} \leq C(N)(|u_n|^{2^*} + |u|^{2^*}) \leq C(N)(|\underline{u}| + |\overline{u}|)^{2^*} \in L^1(\Omega)\,.$$

Lebesgue's theorem implies that $u_n \to u$ in $L^{2^*}(\Omega)$. Using Theorem 3.6 and hypothesis 2 on g, $g(u_n) \to g(u)$ in $L^{\frac{2N}{N+2}}(\Omega)$.

Let us prove that u is a solution to problem (4.5.1). By the definition of u_n, one has

$$\int_\Omega M(x)\nabla u_n \cdot \nabla u_n = \int_\Omega (g(u_{n-1}) + f)\,u_n\,.$$

Using Hölder's inequality on the right-hand side and the ellipticity of M on the left-hand side, one gets

$$\alpha\|\nabla u_n\|^2_{L^2(\Omega)} \leq \|g(u_{n-1}) + f\|_{L^{\frac{2N}{N+2}}(\Omega)}\|u_n\|_{L^{2^*}(\Omega)}\,.$$

The right-hand side is uniformly bounded; therefore u_n is bounded in $H_0^1(\Omega)$ and, up to a subsequence, has a weak limit in $H_0^1(\Omega)$, which necessarily is u. We can now pass to the limit in

$$\int_\Omega M(x)\nabla u_{n+1} \cdot \nabla v = \int_\Omega [g(u_n) + f]\, v\,, \quad \forall\, v \in H_0^1(\Omega)$$

to get

$$\int_\Omega M(x)\nabla u \cdot \nabla v = \int_\Omega [g(u) + f]\, v\,, \quad \forall\, v \in H_0^1(\Omega)$$

that is, u is a solution to problem (4.5.1). □

Example 4.13. Let f be a positive $L^\infty(\Omega)$ function and $\Omega \subset \mathbb{R}^N, N \le 6$. Using the previous theorem it is easy to prove that the following problem:

$$\begin{cases} -\Delta u = u^2 - f\,, & \text{in } \Omega\,, \\ u = 0\,, & \text{on } \partial\Omega\,, \end{cases} \tag{4.5.2}$$

has a solution. Indeed $\underline{u} = 0$ is a supersolution. On the other hand, using Theorem 4.12 the solution ψ to problem

$$\begin{cases} -\Delta \psi = -f\,, & \text{in } \Omega\,, \\ \psi = 0\,, & \text{on } \partial\Omega\,, \end{cases}$$

is a negative subsolution to problem (4.5.2).

4.6 Appendix

We recall here the results on Hilbert spaces and Sobolev spaces that we have used in this chapter (we refer to [22] for more details).

4.6.1 A brief review of functional analysis

Definition 4.14. Let H be a Hilbert space. Let $a : H \times H \to \mathbb{R}$ be a bilinear form.

(1) a is continuous if there exists $\beta > 0$ such that

$$|a(u,v)| \le \beta \|u\| \|v\|\,, \quad \forall\, u, v \in H\,.$$

(2) a is coercive if there exists $\alpha > 0$ such that

$$a(u,u) \ge \alpha \|u\|^2\,, \quad \forall\, u \in H\,.$$

Theorem 4.15 (Projection). *Let H be a Hilbert space and let $K \subset H$ be a closed, not empty, convex set. Then for every g in H there exists a unique point in K, denoted by $P_K g$, such that*

$$\|g - P_K g\| \le \|g - v\|\,, \quad \forall\, v \in K\,.$$

Moreover, $P_K g$ satisfies

$$(g - P_K g | v - P_K g) \le 0\,, \quad \forall\, v \in K \tag{4.6.1}$$

and

$$\|P_K g_1 - P_K g_2\| \le \|g_1 - g_2\|\,, \quad \forall\, g_1, g_2 \in K\,. \tag{4.6.2}$$

Theorem 4.16 (Riesz). *Let H be a Hilbert space and let $\varphi : H \to \mathbb{R}$ be a continuous linear functional. Then there exists a unique $g \in H$ such that*

$$\langle \varphi, v \rangle = (g|v)\,, \quad \forall\, v \in H\,.$$

4.6.2 A brief review on Sobolev spaces

Let $1 \le p < N$; we will denote by p^* the real number such that

$$\frac{1}{p^*} = \frac{1}{p} - \frac{1}{N}\,.$$

Theorem 4.17 (Sobolev embeddings). *The following embeddings are continuous:*
(1) $W^{1,p}(\Omega) \subset L^{p^*}(\Omega)$ *if* $1 \le p < N$;
(2) $W^{1,p}(\Omega) \subset L^q(\Omega)$, $\forall\, q \in [p, +\infty)$ *if* $p = N$;
(3) $W^{1,p}(\Omega) \subset L^\infty(\Omega)$ *if* $p > N$.

In particular, for every $u \in W_0^{1,p}(\Omega)$, there exists a positive constant S depending only on N and p such that

$$S\|u\|_{L^{p^*}(\Omega)} \le \|\nabla u\|_{L^p(\Omega)};$$

this inequality is called Sobolev's inequality.

Theorem 4.18 (Rellich–Kondrachov). *The following embeddings are compact:*
(1) $W^{1,p}(\Omega) \subset L^q(\Omega)$, $\forall q \in [1, p^*)$ *if* $1 \le p < N$;
(2) $W^{1,p}(\Omega) \subset L^q(\Omega)$, $\forall q \in [1, +\infty)$ *if* $p = N$;
(3) $W^{1,p}(\Omega) \subset C(\overline{\Omega})$ *if* $p > N$.

In particular, $W^{1,p}(\Omega) \subset L^p(\Omega)$ for every p and the embedding is compact.

Remark 4.19. We recall that if X and Y are Banach spaces, an operator $T : X \to Y$ is compact if the image of a bounded subset of X is relatively compact in Y.

Theorem 4.20. *Let $1 < p < +\infty$. Then $W_0^{1,p}(\Omega)$ is reflexive, that is, the unit ball is compact for the weak topology in $W_0^{1,p}(\Omega)$.*

Theorem 4.21 (Poincaré's inequality). *Let $1 \le p < +\infty$. Then there exists a positive constant $c = c(\Omega, p)$ such that*

$$\|u\|_{L^p(\Omega)} \le c\|\nabla u\|_{L^p(\Omega)}, \quad \forall u \in W_0^{1,p}(\Omega).$$

In particular, $\|\nabla u\|_{(L^p(\Omega))^N}$ is a norm on $W_0^{1,p}(\Omega)$ which is equivalent to the norm $\|u\|_{W_0^{1,p}(\Omega)}$.

Corollary 4.22. *Let $1 < p < +\infty$. Let u_n be a bounded sequence of $W_0^{1,p}(\Omega)$ functions. Then there exist a subsequence and a $W_0^{1,p}(\Omega)$ function u such that $\nabla u_n \to \nabla u$ weakly in $L^p(\Omega)$.*

Theorem 4.23. *Let $F \in W^{-1,p'}(\Omega)$, $p \in (1, \infty)$. Then there exist $f_0, f_1, \dots, f_N \in L^{p'}(\Omega)$ such that*

$$\langle F, v\rangle = \int_\Omega f_0\, v + \sum_{i=1}^{N} \int_\Omega f_i \frac{\partial v}{\partial x_i}, \qquad \forall v \in W_0^{1,p}(\Omega).$$

5 Nonlinear elliptic equations

5.1 Introduction

In this chapter, we prove an existence result for solutions to certain nonlinear elliptic problems. More precisely, we study the following boundary value problem:

$$(LL)\begin{cases} -\mathrm{div}(a(x,u,\nabla u)) = F(x,u,\nabla u)\,, & \text{in} \quad \Omega\,,\\ u = 0\,, & \text{on} \quad \partial\Omega\,. \end{cases}$$

The existence of solutions u was proved by Leray and Lions in [34]. This is the reason why we call (LL) Leray–Lions problem.

Theorem 5.1 (Leray–Lions). *Let Ω be an open bounded set of $\mathbb{R}^N$ and $p \in (1,\infty)$. Let $a : \Omega \times \mathbb{R} \times \mathbb{R}^N \to \mathbb{R}^N$ and $F : \Omega \times \mathbb{R} \times \mathbb{R}^N \to \mathbb{R}$ be two Carathéodory functions, with the following properties:*

(1) *there exists $\beta > 0$ such that $|a(x,s,\xi)| \le \beta\,[|s|^{p-1} + |\xi|^{p-1}]$;*
(2) *there exists $\alpha > 0$ such that $a(x,s,\xi)\cdot\xi \ge \alpha|\xi|^p\,, \quad \forall\,\xi \in \mathbb{R}^N$;*
(3) *$[a(x,s,\xi) - a(x,s,\eta)]\cdot[\xi-\eta] > 0$ if $\xi \ne \eta$;*
(4) *there exists $f \in L^{p'}(\Omega)$ such that $|F(x,s,\xi)| \le f(x)$.*

Then there exists a solution $u \in W_0^{1,p}(\Omega)$ to problem (LL), that is,

$$\int_\Omega a(x,u,\nabla u)\cdot\nabla v = \int_\Omega F(x,u,\nabla u)\,v\,, \quad \forall\,v \in W_0^{1,p'}(\Omega)\,.$$

We point out that the proof of this theorem is based on an abstract result of surjectivity. Related results proved by Brezis, Browder, and Minty can be found in [21], [25], and [39], respectively.

As we will see, the nonlinearity a makes the problem much more difficult than the semilinear problems of Chapter 4.

5.2 Surjectivity theorem

We will need the following definitions. If V is a Banach space, $||x||$ will denote the norm of an element $x \in V$, $\|\varphi\|_{V'}$ the norm of an element $\varphi \in V'$ and finally $\langle\varphi, v\rangle$ will denote $\varphi(v)$, for $\varphi \in V'$ and $v \in V$.

Definition 5.2. Let V be a Banach reflexive space. An operator $A : V \to V'$ is pseudomonotone if

(1) A is bounded, that is, the image of a bounded subset of V is a bounded subset of V';

(2) if $u_j \to u$ weakly in V and if $\limsup_{j \to +\infty} \langle A(u_j), u_j - u \rangle \leq 0$, then

$$\liminf_{j \to +\infty} \langle A(u_j), u_j - v \rangle \geq \langle A(u), u - v \rangle$$

for every v in V.

Definition 5.3. Let V be a Banach reflexive space. An operator $A : V \to V'$ is coercive if

$$\frac{\langle A(v), v \rangle}{\|v\|} \to +\infty, \quad \text{as} \quad \|v\| \to +\infty .$$

In the proof of the surjectivity theorem we will use the following lemma:

Lemma 5.4. *Let $T : \mathbb{R}^m \to \mathbb{R}^m$ be a continuous map, $m \geq 1$. Assume that there exists $\rho > 0$ such that $T(\xi) \cdot \xi \geq 0$, for every ξ with $|\xi| = \rho$. Then there exists ξ with $|\xi| \leq \rho$ such that $T(\xi) = 0$.*

Proof. By contradiction, assume that $T(\xi) \neq 0$ in $\overline{B(0,\rho)} = \{\xi \in \mathbb{R}^m : |\xi| \leq \rho\}$. We consider the following continuous map from $\overline{B(0,\rho)}$ to itself:

$$\xi \to \frac{-T(\xi)\rho}{|T(\xi)|} .$$

Brouwer's theorem (Theorem 2.4) implies that there exists a fixed point, that is, there exists ξ such that

$$\xi = -\frac{T(\xi)\rho}{|T(\xi)|} .$$

We deduce that $|\xi| = \rho$. On the other hand, $T(\xi) \cdot \xi = -\rho|T(\xi)| < 0$. This is in contradiction with the hypothesis that $T(\xi) \cdot \xi \geq 0$ for every ξ such that $|\xi| = \rho$. □

We can now prove the surjectivity theorem.

Theorem 5.5 (Surjectivity Theorem). *Let V be a Banach reflexive, separable space. Let $A : V \to V'$ be a pseudomonotone coercive operator. Then A is surjective, that is, for every f in V' there exists u in V such that $A(u) = f$.*

Proof. We will divide the proof into several steps.

Step I: Let $\{w_1, \ldots w_n, \ldots\}$ be a dense and countable subset of V. Let us denote with V_n the subspace of V generated by $\{w_1, \ldots w_n\}$. We define

$$\begin{aligned} T : \quad & V_n \to V_n \\ & v \to < A(v) - f, v > v . \end{aligned}$$

We are going to prove that T is continuous. It suffices to prove that $v \to \langle A(v), v\rangle$ is continuous on V_n. Assume that $w_m \to w$: we claim that $A(w_m) \to A(w)$ weakly in V'. Since A is bounded, $\|A(w_m)\|_{V'}$ is bounded uniformly. This implies that

$$\lim_{m\to\infty} \langle A(w_m), w_m - w\rangle = 0\,.$$

Moreover there exists a subsequence such that $A(w_{m_k}) \to g$ weakly in V', as V' is reflexive. This and the pseudomonotonicity imply that

$$\liminf_{m_k\to\infty} \langle A(w_{m_k}), w_{m_k} - v\rangle = \langle g, w - v\rangle \geq \langle A(w), w - v\rangle$$

for every $v \in V$. Then $g = A(w)$ and therefore $A(w_{m_k}) \to A(w)$ weakly in V'.

Now, assume that $A(w_m)$ does not converge weakly to $A(w)$ in V': the previous argument implies a contradiction. Since $A(w_m) \to A(w)$ weakly in V' and $w_m \to w$ in V, then $\langle A(w_m), w_m\rangle \to \langle A(w), w\rangle$. This implies the continuity of T.

Step II: We are going to prove that for every $n \in \mathbb{N}$ there exists $u_n \in V_n$ such that

$$\langle A(u_n), w_j\rangle = \langle f, w_j\rangle\,, \quad 1 \leq j \leq n\,. \tag{5.2.1}$$

We claim that T satisfies $T(v)\cdot v \geq 0$ for every v with $\|v\| = \rho$, for a certain $\rho > 0$. Since T is continuous by the previous step, Lemma 5.4 will imply the existence of u_n.

We have that $T(v)\cdot v \geq 0$, because

$$\langle A(v), v\rangle - \langle f, v\rangle \geq \langle A(v), v\rangle - \|f\|_{V'}\|v\| \geq 0\,,$$

if $\|v\| = \rho$, for ρ sufficiently large, due to the coercivity of A.

Step III: From (5.2.1), we deduce that

$$\langle A(u_n), u_n\rangle = \langle f, u_n\rangle \leq \|f\|_{V'}\,\|u_n\|\,.$$

Using the coercivity of A, we found that $\|u_n\|$ is uniformly bounded. Since A is bounded, $\|A(u_n)\|_{V'}$ is uniformly bounded too. Consequently there exists a subsequence u_{n_k} such that

$$\begin{cases} u_{n_k} \to u & \text{weakly in } V\,,\\ A(u_{n_k}) \to \chi & \text{weakly in } V'\,. \end{cases} \tag{5.2.2}$$

Passing to the limit for $n \to +\infty$ in (5.2.1), (with j fixed) we have for every j

$$\langle \chi, w_j\rangle = \langle f, w_j\rangle\,.$$

Since $\{w_1, \ldots, w_n, \ldots\}$ is a dense set of V, $\chi = f$. Let us prove that

$$\chi = A(u)\,.$$

We claim that

$$\limsup_{n_k \to \infty} \langle A(u_{n_k}), u_{n_k} - u \rangle \le 0 \,. \tag{5.2.3}$$

For a given $\varepsilon > 0$, there exists $u_0 \in \cup_m V_m$ such that $||u - u_0|| \le \varepsilon$. Now,

$$\langle A(u_{n_k}), u_{n_k} - u \rangle = \langle f, u_{n_k} \rangle - \langle A(u_{n_k}), u - u_0 \rangle - \langle A(u_{n_k}), u_0 \rangle .$$

By (5.2.2) the first term of the right-hand side tends to $\langle f, u \rangle$ and the third one tends to $\langle \chi, u_0 \rangle = \langle f, u_0 \rangle$; the second one can be estimated by $C\varepsilon$, for some $C > 0$, since A is bounded. Using that $||u - u_0|| \le \varepsilon$ we get (5.2.3). The pseudomonotonicity of A implies that

$$\langle A(u), u - v \rangle \le \liminf_{n_k \to \infty} \langle A(u_{n_k}), u_{n_k} - v \rangle \le \langle \chi, u - v \rangle$$

for every $v \in V$. It follows that $\chi = A(u)$, that is, $f = A(u)$. □

5.3 The Leray–Lions existence theorem

We are going to prove the Leray–Lions theorem. We recall the statement

Theorem 5.6. *Let Ω be an open bounded set of $\mathbb{R}^N$ and $p \in (1, \infty)$. Let $a : \Omega \times \mathbb{R} \times \mathbb{R}^N \to \mathbb{R}^N$ and $F : \Omega \times \mathbb{R} \times \mathbb{R}^N \to \mathbb{R}$ be two Carathéodory functions, with the following properties:*

(1) *there exists $\beta > 0$ such that $|a(x, s, \xi)| \le \beta\,[|s|^{p-1} + |\xi|^{p-1}]$;*
(2) *there exists $\alpha > 0$ such that $a(x, s, \xi) \cdot \xi \ge \alpha |\xi|^p, \quad \forall\, \xi \in \mathbb{R}^N$;*
(3) *$[a(x, s, \xi) - a(x, s, \eta)] \cdot [\xi - \eta] > 0$ if $\xi \ne \eta$;*
(4) *there exists $f \in L^{p'}(\Omega)$ such that $|F(x, s, \xi)| \le f(x)$.*

Then there exists a solution $u \in W_0^{1,p}(\Omega)$ to problem (LL), that is,

$$\int_\Omega a(x, u, \nabla u) \cdot \nabla v = \int_\Omega F(x, u, \nabla u)\, v \,, \quad \forall\, v \in W_0^{1,p'}(\Omega) \,.$$

Remark 5.7. An application a satisfying hypothesis 2 will be called elliptic.

In the Leray–Lions theorem we will apply the surjectivity theorem to the operator

$$\begin{aligned} A:\ \ W_0^{1,p}(\Omega) &\to W^{-1,p}(\Omega) \\ v &\to -\mathrm{div}(a(x, v, \nabla v)) - F(x, v, \nabla v) \,. \end{aligned}$$

We will use the following lemma to prove that A is pseudomonotone.

Lemma 5.8. *Assume that $u_n \to u$ weakly in $W_0^{1,p}(\Omega)$ and that $[a(x,u_n,\nabla u_n) - a(x,u,\nabla u)] \cdot \nabla(u_n - u) \to 0$ a.e. in Ω. Then $\nabla u_n \to \nabla u$ a.e. in Ω.*

Proof. Since

$$[a(x,u_n,\nabla u_n) - a(x,u,\nabla u)] \cdot \nabla(u_n - u) \to 0 \quad \text{a.e. in} \quad \Omega$$

one has

$$|[a(x,u_{n_k},\nabla u_{n_k}) - a(x,u,\nabla u)] \cdot \nabla(u_{n_k} - u)| \le C(x)$$

for some function $C(x)$. Up to a Lebesgue measure zero set Z, the above inequality holds pointwise. Let us prove that there exists a function c such that

$$|\nabla u_{n_k}(x)| \le c(x)\,. \tag{5.3.1}$$

One has, by hypotheses 1 and 2 on a

$$\begin{aligned} C(x) &\ge [a(x,u_{n_k},\nabla u_{n_k}) - a(x,u,\nabla u)] \cdot \nabla(u_{n_k} - u) \\ &\ge \alpha[|\nabla u_{n_k}|^p + |\nabla u|^p] - |\nabla u_{n_k}|[\beta(|u|^{p-1} + |\nabla u|^{p-1})] \\ &\quad - |\nabla u|[\beta(|u_{n_k}|^{p-1} + |\nabla u_{n_k}^{p-1}|)]\,. \end{aligned} \tag{5.3.2}$$

The $W_0^{1,p}(\Omega)$ weak convergence of u_n to u implies the existence of a subsequence (still denoted by u_n) and of a function g in $L^1(\Omega)$ such that

$$|u_n|^{p-1}|\nabla u| \le g \quad \text{and} \quad u_n \to u \quad \text{a.e. in} \quad \Omega.$$

Since in (5.3.2) we have a polynomial in $|\nabla u_n|$, (5.3.1) follows. We are going to prove that

$$\nabla u_n(x) \to \nabla u(x) \quad \text{in} \quad \Omega \setminus Z\,. \tag{5.3.3}$$

Assume by contradiction that there exists $x_0 \in \Omega \setminus Z$ such that $\nabla u_n(x_0)$ does not converge to $\nabla u(x_0)$. The Bolzano–Weierstrass theorem implies that $\nabla u_n(x_0) \to b$, for some $b \in \mathbb{R}^N$, up to a subsequence. Passing to the limit in

$$[a(x_0,u_n(x_0),\nabla u_n(x_0)) - a(x_0,u(x_0),\nabla u(x_0))] \cdot \nabla(u_n(x_0) - u(x_0))$$

we get

$$[a(x_0,u(x_0),b) - a(x_0,u(x_0),\nabla u(x_0))] \cdot (b - \nabla u(x_0)) = 0$$

which yields $b = \nabla u(x_0)$ by hypothesis 3 on a. □

Lemma 5.9. *Let u_n, u be $W_0^{1,p}(\Omega)$ functions such that $u_n \to u$ weakly in $W_0^{1,p}(\Omega)$. If*

$$\int_\Omega [a(x,u_n,\nabla u_n) - a(x,u,\nabla u)] \cdot \nabla(u_n - u) \to 0\,, \tag{5.3.4}$$

then $a(x,u_n,\nabla u_n) \to a(x,u,\nabla u)$ weakly in $L^{p'}(\Omega)$.

Proof. We claim that

$$[a(x,u_n,\nabla u_n)-a(x,u,\nabla u)]\cdot\nabla(u_n-u)\to 0\quad\text{in}\quad L^1(\Omega)\,. \tag{5.3.5}$$

We can write

$$[a(x,u_n,\nabla u_n)-a(x,u,\nabla u)]\cdot\nabla(u_n-u)$$

as the sum of

$$\alpha_n=[a(x,u_n,\nabla u_n)-a(x,u_n,\nabla u)]\cdot\nabla(u_n-u)$$

and

$$\beta_n=[a(x,u_n,\nabla u)-a(x,u,\nabla u)]\cdot\nabla(u_n-u)\,.$$

We observe that $\beta_n\to 0$ in $L^1(\Omega)$, since by Hölder's inequality

$$\int_\Omega|\beta_n|\le\|a(x,u_n,\nabla u)-a(x,u,\nabla u)\|_{L^{p'}(\Omega)}\|\nabla u_n-\nabla u\|_{L^p(\Omega)}\,.$$

Now, $\|\nabla u_n-\nabla u\|_{L^p(\Omega)}$ is bounded, and $a(x,u_n,\nabla u)\to a(x,u,\nabla u)$ in $L^{p'}(\Omega)$ due to Theorem 3.6, as $u_n\to u$ in $L^p(\Omega)$. By (5.3.4) $\int_\Omega\alpha_n+\beta_n\to 0$ and $\beta_n\to 0$ in $L^1(\Omega)$ as we have just proved. This implies that $\int_\Omega\alpha_n\to 0$; since $\alpha_n\ge 0$ for the monotonicity of a, we have that $\alpha_n+\beta_n\to 0$ in $L^1(\Omega)$, that is, (5.3.5) holds.

Up to a subsequence,

$$[a(x,u_n,\nabla u_n)-a(x,u,\nabla u)]\cdot\nabla(u_n-u)\to 0$$

a.e. in Ω. By Lemma 5.8 $\nabla u_n\to\nabla u$ a.e. in Ω. Theorem 3.7 implies the result. □

We can now prove the Leray–Lions theorem.

Proof. We will prove that $A(v)=-\operatorname{div}(a(x,v,\nabla v))-F(x,v,\nabla v)$ is coercive and pseudomonotone; the result will follow from Theorem 5.5.

Step I: Hypotheses 2 and 4 on a give

$$\langle A(v),v\rangle\ge\alpha\int_\Omega|\nabla v|^p-\int_\Omega|f||v|\ge\alpha\int_\Omega|\nabla v|^p-\|f\|_{L^{p'}(\Omega)}\|v\|_{L^p(\Omega)}\,.$$

Poincaré's inequality implies the coercivity of A.

Step II: By hypotheses 1 and 4 on a one has

$$\begin{aligned}\langle A(v),w\rangle&=\int_\Omega a(x,v,\nabla v)\cdot\nabla w-\int_\Omega F(x,v,\nabla v)w\\&\le\beta\left[\int_\Omega|v|^{p-1}|\nabla w|+\int_\Omega|\nabla v|^{p-1}|\nabla w|\right]+\int_\Omega|f||w|\\&\le\beta\left[\|v\|^{p-1}_{L^p(\Omega)}\|\nabla w\|_{L^p(\Omega)}+\|\nabla v\|^{p-1}_{L^p(\Omega)}\|\nabla w\|_{L^p(\Omega)}\right.\\&\qquad\left.+\|f\|_{L^{p'}(\Omega)}\|w\|_{L^p(\Omega)}\right]\end{aligned}$$

and so $\|A(v)\|_{W^{-1,p'}(\Omega)}$ is bounded, if $\|v\|_{W_0^{1,p}(\Omega)}$ is bounded.

Step III: Assume that $u_n \to u$ weakly in $W_0^{1,p}(\Omega)$ and

$$\begin{aligned} 0 &\geq \limsup_{n\to+\infty} \langle A(u_n), u_n - u\rangle \\ &= \limsup_{n\to+\infty} \int_\Omega a(x,u_n,\nabla u_n)\cdot\nabla(u_n-u) + \int_\Omega F(x,u_n,\nabla u_n)(u_n-u)\,. \end{aligned} \tag{5.3.6}$$

We are going to prove that

$$\lim_{n\to+\infty} \int_\Omega [a(x,u_n,\nabla u_n) - a(x,u,\nabla u)]\cdot\nabla(u_n-u) = 0\,. \tag{5.3.7}$$

This will be used in *Step IV* to prove the pseudomonotonicity of the operator.

We observe that $\int_\Omega F(x,u_n,\nabla u_n)(u_n-u) \to 0$, since $u_n \to u$ in $L^p(\Omega)$ and the sequence $F(x,u_n,\nabla u_n)$ is bounded in $L^{p'}(\Omega)$ due to the hypotheses on F. By (5.3.6) this implies that

$$\limsup_{n\to+\infty} \int_\Omega [a(x,u_n,\nabla u_n) - a(x,u,\nabla u)]\cdot\nabla(u_n-u) \leq 0\,.$$

Using hypothesis 3 on a, we have

$$\begin{aligned} &\int_\Omega [a(x,u_n,\nabla u_n) - a(x,u,\nabla u)]\cdot\nabla(u_n-u) \\ &\geq \int_\Omega [a(x,u_n,\nabla u) - a(x,u,\nabla u)]\cdot\nabla(u_n-u) \end{aligned}$$

and this last term goes to 0, since $a(x,u_n,\nabla u) \to a(x,u,\nabla u)$ by Theorem 3.6. We have thus proved (5.3.7).

Step IV: Assume that $u_n \to u$ weakly in $W_0^{1,p}(\Omega)$ and (5.3.6). We will prove that for every $w \in W_0^{1,p}(\Omega)$

$$\liminf_{n\to+\infty} \langle A(u_n), u_n - w\rangle \geq \langle A(u), u - w\rangle\,. \tag{5.3.8}$$

We remark that

$$\langle A(u_n), u_n - w\rangle = \int_\Omega a(x,u_n,\nabla u_n)\cdot\nabla(u_n-w) + \int_\Omega F(x,u_n,\nabla u_n)(u_n-w)\,.$$

We will separately study the two terms of the right-hand side.

The limit of (5.3.7) allows us to apply Lemma 5.9 to deduce that $a(x,u_n,\nabla u_n) \to a(x,u,\nabla u)$ weakly in $L^{p'}(\Omega)$, and so

$$\int_\Omega a(x,u_n,\nabla u_n)\cdot\nabla w \to \int_\Omega a(x,u,\nabla u)\cdot\nabla w \tag{5.3.9}$$

for every $w \in W_0^{1,p}(\Omega)$. On the other hand, by Lemma 5.8, $a(x,u_n,\nabla u_n) \to a(x,u,\nabla u)$ and $\nabla u_n \to \nabla u$ a.e. in Ω. Since $a(x,u_n,\nabla u_n)\cdot\nabla u_n \geq 0$ by the ellipticity of a, Fatou's Lemma implies that

$$\liminf_{n\to+\infty}\int_\Omega a(x,u_n,\nabla u_n)\cdot\nabla u_n \geq \int_\Omega a(x,u,\nabla u)\cdot\nabla u. \tag{5.3.10}$$

From (5.3.9) and (5.3.10), we deduce that

$$\liminf_{n\to+\infty}\int_\Omega a(x,u_n,\nabla u_n)\cdot\nabla(u_n-w) \geq \int_\Omega a(x,u,\nabla u)\cdot\nabla(u-w).$$

Let us finally study

$$\int_\Omega F(x,u_n,\nabla u_n)(u_n-w).$$

Since $u_n \to u$ and $\nabla u_n \to \nabla u$ a.e. in Ω and weakly in $L^p(\Omega)$, using Theorem 3.7 one has that $F(x,u_n,\nabla u_n) \to F(x,u,\nabla u)$ weakly in $L^{p'}(\Omega)$. This implies that

$$\int_\Omega F(x,u_n,\nabla u_n)(u_n-w) \to \int_\Omega F(x,u,\nabla u)(u-w)$$

since $u_n \to u$ in $L^p(\Omega)$.

We thus have obtained

$$\begin{aligned}\liminf_{n\to+\infty}\langle A(u_n),u_n-w\rangle &\geq \int_\Omega a(x,u,\nabla u)\cdot\nabla(u-w)+\int_\Omega F(x,u,\nabla u)(u-w)\\ &= \langle A(u),u-w\rangle.\end{aligned}$$

Therefore A is pseudomonotone. □

We observe that in the following chapters we will essentially use the case $p=2$, that is, the following corollary:

Corollary 5.10. *Let $a:\Omega\times\mathbb{R}\times\mathbb{R}^N \to \mathbb{R}^N$ be a Carathéodory function, with the following properties:*
(1) *there exists $\beta>0$ such that $|a(x,s,\xi)| \leq \beta\,[|s|+|\xi|]$;*
(2) *there exists $\alpha>0$ such that $a(x,s,\xi)\cdot\xi \geq \alpha|\xi|^2$;*
(3) *$[a(x,s,\xi)-a(x,s,\eta)]\cdot[\xi-\eta]>0$ if $\xi\neq\eta$.*

Then the operator

$$A: v \to -div(a(x,v,\nabla v)),$$

defined on $H_0^1(\Omega)$ into $H^{-1}(\Omega)$, is surjective. In particular, if $f\in L^m(\Omega)$, with $m\geq\frac{2N}{N+2}$, or $f\in M^m(\Omega)$, with $m>\frac{2N}{N+2}$, there exists $u\in H_0^1(\Omega)$ solution to $-div(a(x,u,\nabla u))=f$.

We will also study elliptic problems with lower order terms, that is,

$$\begin{cases} -\mathrm{div}(a(x,u,\nabla u)) + u = f, & \text{in } \Omega, \\ u = 0, & \text{on } \partial\Omega. \end{cases}$$

Theorem 5.11. *Let Ω be an open bounded set of $\mathbb{R}^N$ and $p \in (1,\infty)$. Let $a : \Omega \times \mathbb{R} \times \mathbb{R}^N \to \mathbb{R}^N$ be a Carathéodory function, with the following properties:*
(1) *there exists $\beta > 0$ such that $|a(x,s,\xi)| \le \beta\,[|s|^{p-1} + |\xi|^{p-1}]$;*
(2) *there exists $\alpha > 0$ such that $a(x,s,\xi)\cdot\xi \ge \alpha|\xi|^p, \quad \forall\, \xi \in \mathbb{R}^N$;*
(3) *$[a(x,s,\xi) - a(x,s,\eta)]\cdot[\xi - \eta] > 0$ if $\xi \ne \eta$.*

Let $f \in L^m(\Omega)$, with $m \ge \frac{2N}{N+2}$. Then there exists a solution $u \in H_0^1(\Omega)$ to problem $-\mathrm{div}\,(a(x,u,\nabla u)) + u = f$.

Proof. We are going to prove that

$$\begin{aligned} A:\quad & H_0^1(\Omega) \to H^{-1}(\Omega) \\ & \qquad v \to -\mathrm{div}(a(x,v,\nabla v)) + v \end{aligned}$$

is pseudomonotone, in order to use Theorem 5.5. In the proof of the Leray–Lions theorem we proved that $v \to -\mathrm{div}(a(x,v,\nabla v))$ is a pseudomonotone coercive operator from $H_0^1(\Omega)$ to $H^{-1}(\Omega)$. For the first point of the definition of a pseudomonotone operator (see Definition 5.2), we have only to prove that $v \to v$ is a bounded operator from $H_0^1(\Omega)$ to $H^{-1}(\Omega)$. Hölder's and Sobolev's inequalities imply that

$$\langle v, w\rangle = \int_\Omega vw \le \|v\|_{L^{\frac{2N}{N+2}}(\Omega)} \|w\|_{L^{\frac{2N}{N-2}}(\Omega)} \le \frac{1}{S}\|v\|_{L^{\frac{2N}{N+2}}(\Omega)} \|w\|_{H_0^1(\Omega)}$$

and so $\|A(v)\|_{H^{-1}(\Omega)}$ is bounded, if $\|v\|_{H_0^1(\Omega)}$ is bounded. Moreover, to prove point 2 of Definition 5.2, we only have to show that if $u_j \to u$ weakly in $H_0^1(\Omega)$, as $j \to \infty$, and if $\limsup_{j\to+\infty} \int_\Omega u_j(u_j - u) \le 0$, then $\liminf_{j\to+\infty} \int_\Omega u_j(u_j - v) \ge \int_\Omega u(u - v)$ for every v in $H_0^1(\Omega)$. This is clear, since

$$\int_\Omega u_j(u_j - v) - u(u - v) = \int_\Omega u_j^2 - u^2 - v(u_j - u) \to 0,$$

as $u_j \to u$ weakly in $H_0^1(\Omega)$ and $u_j \to u$ in $L^2(\Omega)$. Finally, A is coercive, since $v \to -\mathrm{div}(a(x,v,\nabla v))$ is coercive, as we have already proved. □

The Leray–Lions theorem can be proved under less strict hypotheses on a and F:

Theorem 5.12. *Let Ω be an open bounded set of $\mathbb{R}^N$ and $p \in (1,\infty)$. Let $a : \Omega \times \mathbb{R} \times \mathbb{R}^N \to \mathbb{R}^N$ and $F : \Omega \times \mathbb{R} \times \mathbb{R}^N \to \mathbb{R}$ be two Carathéodory functions, with the following*

properties:
(1) *there exists* $\beta_1 > 0$ *such that* $|a(x,s,\xi)| \le \beta_1 [|s|^{\frac{p^*-\varepsilon}{p'}} + |\xi|^{p-1}]$, *for some* $\varepsilon > 0$;
(2) *there exists* $\alpha > 0$ *such that* $a(x,s,\xi) \cdot \xi \ge \alpha |\xi|^p, \quad \forall\, \xi \in \mathbb{R}^N$;
(3) $[a(x,s,\xi) - a(x,s,\eta)] \cdot [\xi - \eta] > 0$ *if* $\xi \ne \eta$;
(4) *there exist* $k_1 \in L^{p^{*'}+\varepsilon}(\Omega)$ *and* $\beta_2 > 0$ *such that* $|F(x,s,\xi)| \le k_1(x) + \beta_2 [|s|^{p^*-\varepsilon-1} + |\xi|^{\frac{p-\varepsilon}{p^{*'}}}]$;
(5) *there exist* $k_2 \in L^1(\Omega)$ *and* $\beta_2 > 0$ *such that* $F(x,s,\xi)s \ge \beta_3 |s|^q - k_2(x), q \le p$.

Then for every $g \in L^{p^{*'}}(\Omega)$ *there exists a function* $u \in W_0^{1,p}(\Omega)$ *such that*

$$\int_\Omega a(x,u,\nabla u) \cdot \nabla v + \int_\Omega F(x,u,\nabla u) v = \int_\Omega g v\,, \quad \forall\, v \in W_0^{1,p'}(\Omega)\,.$$

6 Summability of the solutions

6.1 Introduction

In the previous chapter, we have proved the existence of weak solutions to the Leray–Lions problem:

$$\begin{cases} -\mathrm{div}(a(x,u,\nabla u)) = f\,, & \text{in} \quad \Omega\,,\\ u = 0\,, & \text{on} \quad \partial\Omega\,. \end{cases}$$

In this chapter we are going to present some regularity results proved in [17, 49, 50, 52]. We will see that the regularity of the solution depends on the regularity of the source. The starting points are some regularity results by Stampacchia. We will assume that the source is a function f belonging to a Lebesgue space or to a Marcinkiewicz space; we will also treat the case where the source is of divergence type.

We illustrate here the results of this chapter in a schematic way. We recall that for an open bounded subset Ω of $\mathbb{R}^N, N \geq 3$, $a : \Omega \times \mathbb{R} \times \mathbb{R}^N \to \mathbb{R}^N$ is a Carathéodory map with the following properties:

(1) there exists $\beta > 0$ such that $|a(x,s,\xi)| \leq \beta\,[|s| + |\xi|]$;
(2) there exists $\alpha > 0$ such that $a(x,s,\xi)\cdot\xi \geq \alpha|\xi|^2, \quad \forall\,\xi \in \mathbb{R}^N$;
(3) $[a(x,s,\xi) - a(x,s,\eta)]\cdot[\xi - \eta] > 0$ if $\xi \neq \eta$.

We will show the following results:

$$\begin{aligned}
&f \in L^m(\Omega)\,,\ m > N/2 \Rightarrow u \in L^\infty(\Omega)\\
&f \in L^m(\Omega)\,,\ m \in [2N/(N+2), N/2) \Rightarrow u \in L^{m^{**}}(\Omega)\\
&f \in L^{\frac{N}{2}}(\Omega) \Rightarrow e^{\lambda|u|} \in L^{2^*}(\Omega)\,,\ \forall\lambda > 0\\
&f \in M^m(\Omega)\,,\ m > N/2 \Rightarrow u \in L^\infty(\Omega)\\
&f \in M^m(\Omega)\,,\ m \in [2N/(N+2), N/2) \Rightarrow u \in M^{m^{**}}(\Omega)\\
&f \in M^{\frac{N}{2}}(\Omega) \Rightarrow \exists b > 0\colon \int_\Omega e^{b|u|} < \infty\,.
\end{aligned}$$

When the source is the divergence of a vector field $F : \Omega \to \mathbb{R}^N$ such that $\mathrm{div} F \in L^2(\Omega)$, if u is a solution, we will prove that

$$\begin{aligned}
&F \in (L^m(\Omega))^N, m > N \Rightarrow u \in L^\infty(\Omega)\\
&F \in (L^m(\Omega))^N, 2 < m < N \Rightarrow u \in L^{m^*}(\Omega)\,.
\end{aligned}$$

In the case where a satisfies the following coercivity:

$$a(x,s,\xi)\cdot\xi \geq \alpha|\xi|^p\,, \quad \forall\,\xi \in \mathbb{R}^N$$

it is possible to prove similar results with the numbers $(p^*)'$, N/p playing the role of $2N/(N+2), N/2$.

6.2 Preliminaries

In this chapter we will often use the function

$$T_k(s) = \begin{cases} -k\,, & s \le -k\,, \\ s\,, & |s| \le k\,, \\ k\,, & s \ge k\,, \end{cases}$$

for $k > 0$ and $G_k(s) = s - T_k(s)$. Moreover, if $f : \Omega \to \mathbb{R}$ is a measurable function, we will use the following notation:

$$g(k) = \int_\Omega |G_k(f)|$$

and

$$A_k = \{|f| > k\}\,.$$

Lemma 6.1. *Let $f \in L^1(\Omega)$. Then $g(k)$ is differentiable a.e. and $g'(k) = -\mathrm{meas}(A_k)$.*

Proof. Let us prove that

$$\tilde{g}(k) = \int_{\{w-k>0\}} (w-k)$$

is differentiable with respect to k. To this end, let us set

$$A_{k,+} = \{w - k > 0\}\,.$$

The function $\tilde{g}(k)$ is differentiable a.e., since it is monotone. Let us compute its derivative. Let $h \in \mathbb{R}^+$; the incremental rate of $\tilde{g}$ is

$$\begin{aligned}
\frac{\tilde{g}(k+h) - \tilde{g}(k)}{h} &= \frac{1}{h}\left(\int_{A_{k+h,+}} (w-k-h) - \int_{A_{k,+}} (w-k)\right) \\
&= \frac{1}{h}\left(\int_{A_{k+h,+}} -h - \int_{\{k<w\le k+h\}} (w-k)\right) \\
&= -\int_\Omega \chi_{\{w>k+h\}} - \frac{1}{h}\int_{\{k<w\le k+h\}} (w-k)\,.
\end{aligned}$$

We have

$$0 \le \int_{\{k<w\le k+h\}} (w-k) \le \int_{\{k<w\le k+h\}} h$$

and so

$$0 \le \frac{1}{h}\int_{\{k<w\le k+h\}} (w-k) \le \int_\Omega \chi_{\{k<w\le k+h\}},$$

which converges to 0, as $h \to 0^+$. Consequently

$$\lim_{h\to 0} \frac{\tilde{g}(k+h)-\tilde{g}(k)}{h} = -\lim_{h\to 0} \int_\Omega \chi_{\{w>k+h\}} = -\text{meas}(\{w>k\}),$$

that is, $\tilde{g}'(k) = -\text{meas}(A_{k,+})$.

For the general case, it is sufficient to use that

$$g(k) = \int_{\{w-k>0\}} (w-k) + \int_{\{-(w-k)>0\}} -(w-k). \qquad \square$$

The following lemma will be useful to us:

Lemma 6.2. *Let $f \in L^1(\Omega)$ such that $g(k)$ satisfies for every k*

$$g(k) \le C\,\text{meas}(A_k)^\alpha$$

with $\alpha > 1$ and $C > 0$. Then $f \in L^\infty(\Omega)$ and there exists a constant $\gamma = \gamma(\alpha, \Omega)$ such that

$$\|f\|_{L^\infty(\Omega)} \le C\gamma.$$

Proof. Using Lemma 6.1, one has

$$g(k) \le C[-g'(k)]^\alpha$$

that is,

$$g'(k)[g(k)]^{-\frac{1}{\alpha}} \le -\frac{1}{C^{\frac{1}{\alpha}}}. \tag{6.2.1}$$

Integrating this inequality on $(0,k)$ we get

$$-\left(1-\frac{1}{\alpha}\right)\frac{k}{C^{\frac{1}{\alpha}}} \ge g(k)^{1-\frac{1}{\alpha}} - g(0)^{1-\frac{1}{\alpha}} = g(k)^{1-\frac{1}{\alpha}} - \|f\|_{L^1(\Omega)}^{1-\frac{1}{\alpha}}.$$

Consequently, for every $k > 0$,

$$g(k)^{1-\frac{1}{\alpha}} \le \|f\|_{L^1(\Omega)}^{1-\frac{1}{\alpha}} - \left(1-\frac{1}{\alpha}\right)\frac{k}{C^{\frac{1}{\alpha}}}.$$

In particular this inequality holds true for $k_0 = \frac{C^{\frac{1}{\alpha}}\|f\|_{L^1(\Omega)}^{1-\frac{1}{\alpha}}}{1-\frac{1}{\alpha}}$. This implies that $g(k_0) = 0$, that is,

$$|f| \le k_0 = \frac{C^{\frac{1}{\alpha}}\|f\|_{L^1(\Omega)}^{1-\frac{1}{\alpha}}}{1-\frac{1}{\alpha}}.$$

By the Hölder inequality

$$|f| \le \frac{C^{\frac{1}{\alpha}}\|f\|_{L^\infty(\Omega)}^{1-\frac{1}{\alpha}}\,\text{meas}(\Omega)^{1-\frac{1}{\alpha}}}{1-\frac{1}{\alpha}},$$

that is,

$$\|f\|_{L^\infty(\Omega)} \le \left(1-\frac{1}{\alpha}\right)^{-\alpha}\text{meas}(\Omega)^{\alpha-1}C. \qquad \square$$

Remark 6.3. If the inequality

$$g(k) \le C\,\mathrm{meas}(A_k)^{\alpha}$$

holds for every $k \ge h_0$, for some $h_0 \in \mathbb{N}$, then one can slightly modify the above proof to prove that

$$\|f\|_{L^\infty(\Omega)} \le \gamma,$$

where $\gamma = \gamma(C, \alpha, h_0, \Omega)$. Indeed it is sufficient to integrate (6.2.1) on (h_0, k) and to follow the same technique as above.

We will use the following results in the case where the source f in the Leray–Lions problem belongs to $M^{\frac{N}{2}}(\Omega)$.

Proposition 6.4. *Let f be a measurable function defined on Ω and let a be a positive constant. Then $\int_\Omega e^{a|f|} < \infty$ if and only if $\sum_{k=0}^{\infty} e^{ak}\,\mathrm{meas}(A_k) < \infty$.*

Proof. We observe that

$$\sum_{k=0}^{\infty} e^{ak}\mathrm{meas}(A_k) = \sum_{k=0}^{\infty} e^{ak} \sum_{i=k}^{\infty} \mathrm{meas}(B_i) = \sum_{i=0}^{\infty} \mathrm{meas}(B_i) \sum_{k=0}^{i} e^{ak} \tag{6.2.2}$$

if $B_i = \{i < |f| < i+1\}$. Since $t \to e^{at}$ is increasing and continuous

$$\sum_{k=k_0}^{n} e^{ak} \le \int_{k_0}^{n+1} e^{at}\,dt \le \sum_{k=k_0}^{n} e^{a(k+1)}. \tag{6.2.3}$$

We will divide the proof into two parts.

Step I: Assume that $\int_\Omega e^{a|f|} < \infty$; we are going to prove that $\sum_{k=0}^{\infty} e^{ak}\mathrm{meas}(A_k)$ is finite. Using (6.2.2) and the first inequality of (6.2.3), one has

$$\begin{aligned}\sum_{k=0}^{\infty} e^{ak}\,\mathrm{meas}(A_k) &\le \sum_{i=0}^{\infty} \mathrm{meas}(B_i) \int_0^{i+1} e^{at}\,dt \\ &= \frac{e^a}{a} \sum_{i=0}^{\infty} \mathrm{meas}(B_i) \left[e^{ai} - \frac{1}{e^a}\right] \\ &\le \frac{e^a}{a} \sum_{i=0}^{\infty} \int_{B_i} \left[e^{a|f|} - \frac{1}{e^a}\right] \\ &= \frac{e^a}{a} \int_\Omega \left[e^{a|f|} - \frac{1}{e^a}\right] < \infty,\end{aligned}$$

since $\int_\Omega e^{a|f|} < \infty$.

Step II: Assume that $\sum_{k=0}^{\infty} e^{a\,k}\,\text{meas}(A_k) < \infty$; we are going to prove that $\int_\Omega e^{a|f|} < \infty$. Using (6.2.2) and the second inequality of (6.2.3), we have

$$
\begin{aligned}
\sum_{k=0}^{\infty} e^{a\,k}\text{meas}(A_k) &= \sum_{i=0}^{\infty} \text{meas}(B_i) \sum_{k=0}^{i} e^{a\,k} \\
&= \sum_{i=0}^{\infty} \text{meas}(B_i) \sum_{j=-1}^{i-1} e^{a(j+1)} \\
&\geq \sum_{i=0}^{\infty} \text{meas}(B_i) \int_{-1}^{i} e^{at}\,dt \\
&= \sum_{i=0}^{\infty} \text{meas}(B_i) \left[\frac{e^{i\,a} - e^{-a}}{a}\right] \\
&\geq \frac{1}{a} \sum_{i=0}^{\infty} \int_{B_i} [e^{a\,(|f|-1)} - e^{-a}] \\
&= \frac{1}{a} \int_\Omega [e^{a\,(|f|-1)} - e^{-a}]\,,
\end{aligned}
$$

and so $\int_\Omega e^{a|f|} < \infty$. □

Proposition 6.5. *Let f be a measurable function defined on Ω; if there exist $a, c > 0$, $k_0 \in \mathbb{N}$ such that $\text{meas}(A_k) \leq \frac{c}{k\,e^{ak}}$ for every $k \geq k_0$, then $\int_\Omega e^{b|f|} < \infty$ for every $b < a$.*

Proof. Using Proposition 6.4, one has $\int_\Omega e^{b|f|} < \infty$ if and only if

$$\sum_{k=0}^{\infty} e^{bk}\,\text{meas}(A_k) < \infty\,.$$

By hypothesis we have

$$\sum_{k=k_0}^{\infty} e^{bk}\text{meas}(A_k) \leq \sum_{k=k_0}^{\infty} e^{bk}\frac{c}{k\,e^{ak}}\,.$$

The last series is finite due to the choice of b, and so $\sum_{k=0}^{\infty} e^{bk}\text{meas}(A_k)$ is finite. □

6.3 Sources in Lebesgue spaces

We are now going to prove some regularity results of the solutions to

$$
\begin{cases}
-\text{div}(a(x,u,\nabla u)) = f\,, & \text{in } \Omega\,, \\
u = 0\,, & \text{on } \partial\Omega\,,
\end{cases}
\tag{6.3.1}
$$

when the source f belongs $L^m(\Omega)$, with $m \geq \frac{2N}{N+2}$.

Theorem 6.6. *Let $f \in L^m(\Omega)$ with $m > N/2$. Then every solution $u \in H_0^1(\Omega)$ to problem (6.3.1) is bounded; moreover the estimate*

$$\|u\|_{L^\infty(\Omega)} \le C\,\|f\|_{L^m(\Omega)}$$

holds, where C depends on N, α and m.

Proof. Let us take $v = G_k(u)$ as a test function in the weak formulation of (6.3.1). By the ellipticity of a we have

$$\alpha \int_\Omega |\nabla G_k(u)|^2 \le \int_\Omega f G_k(u)\,. \tag{6.3.2}$$

Sobolev's inequality implies that

$$\alpha S^2 \left(\int_\Omega |G_k(u)|^{2^*} \right)^{\frac{2}{2^*}} \le \int_\Omega f G_k(u)\,. \tag{6.3.3}$$

Let us now study the right-hand side: applying the Hölder inequality twice, first with exponent 2^* and then with $m\frac{N+2}{2N}$, we get

$$\begin{aligned} \int_\Omega f G_k(u) &\le \left(\int_\Omega |G_k(u)|^{2^*} \right)^{\frac{1}{2^*}} \left(\int_{A_k} |f|^{\frac{2N}{N+2}} \right)^{\frac{N+2}{2N}} \\ &\le \|f\|_{L^m(\Omega)} \left(\int_\Omega |G_k(u)|^{2^*} \right)^{\frac{1}{2^*}} \operatorname{meas}(A_k)^{\left[1-\frac{2N}{(N+2)m}\right]\frac{N+2}{2N}}\,. \end{aligned} \tag{6.3.4}$$

Estimates (6.3.3) and (6.3.4) thus give

$$\alpha S^2 \left(\int_\Omega |G_k(u)|^{2^*} \right)^{\frac{2}{2^*}} \le \|f\|_{L^m(\Omega)} \left(\int_\Omega |G_k(u)|^{2^*} \right)^{\frac{1}{2^*}} \operatorname{meas}(A_k)^{\left[1-\frac{2N}{(N+2)m}\right]\frac{N+2}{2N}}\,,$$

that is,

$$\alpha S^2 \left(\int_\Omega |G_k(u)|^{2^*} \right)^{\frac{1}{2^*}} \le \|f\|_{L^m(\Omega)} \operatorname{meas}(A_k)^{\left[1-\frac{2N}{(N+2)m}\right]\frac{N+2}{2N}}\,. \tag{6.3.5}$$

By Hölder's inequality with exponent 2^* one has

$$\int_\Omega |G_k(u)| \le \left(\int_\Omega |G_k(u)|^{2^*} \right)^{\frac{1}{2^*}} \operatorname{meas}(A_k)^{\frac{N+2}{2N}}$$

and so (6.3.5) implies that

$$g(k) = \int_\Omega |G_k(u)| \le \frac{1}{\alpha S^2}\|f\|_{L^m(\Omega)} \text{meas}(A_k)^{1+\frac{2}{N}-\frac{1}{m}}.$$

Lemma 6.2 with $\alpha = 1 + \frac{2}{N} - \frac{1}{m}$ and $C = \frac{1}{\alpha S^2}\|f\|_{L^m(\Omega)}$ gives the result. □

By considering the same proof and dropping the positive contribute given by the lower order term, one has the following result:

Theorem 6.7. *Let $f \in L^m(\Omega)$ with $m > N/2$. Then every solution $u \in H_0^1(\Omega)$ to problem*

$$\begin{cases} -\text{div}(a(x,u,\nabla u)) + u = f, & \text{in } \Omega, \\ u = 0, & \text{on } \partial\Omega. \end{cases}$$

is bounded.

Remark 6.8. Observe that if a function u satisfies inequality (6.3.2), then u is bounded.

We can now pass to the regularity of the solutions in the case where $f \in L^m(\Omega)$, with $\frac{2N}{N+2} \le m < \frac{N}{2}$.

Theorem 6.9. *Let $f \in L^m(\Omega)$ with $\frac{2N}{N+2} \le m < \frac{N}{2}$. Then every solution $u \in H_0^1(\Omega)$ to problem* (6.3.1) *belongs to $L^{m^{**}}(\Omega)$; moreover the estimate*

$$\|u\|_{L^{m^{**}}(\Omega)} \le C\,\|f\|_{L^m(\Omega)}$$

holds, where C depends on N, m and α.

Proof. Consider as a test function

$$v = \frac{|T_k(u)|^{2\lambda}T_k(u)}{2\lambda+1},$$

with $\lambda > 0$. We will study separately the two sides of the weak formulation of (6.3.1). For the left-hand side we have

$$\frac{1}{2\lambda+1}\int_\Omega a(x,u,\nabla u)\cdot\nabla(|T_k(u)|^{2\lambda}T_k(u)) \ge \alpha S^2\left(\int_\Omega |T_k(u)|^{(\lambda+1)2^*}\right)^{\frac{2}{2^*}}$$

by using the ellipticity of a and Sobolev's inequality. For the right-hand side, we get

$$\frac{1}{2\lambda+1}\int_\Omega f|T_k(u)|^{2\lambda}T_k(u) \le \frac{1}{2\lambda+1}\left(\int_\Omega |T_k(u)|^{(2\lambda+1)m'}\right)^{\frac{1}{m'}}\|f\|_{L^m(\Omega)}$$

by using Hölder's inequality with exponent m. We have thus proved that

$$\alpha S^2(2\lambda+1)\left[\int_\Omega |T_k(u)|^{(\lambda+1)2^*}\right]^{\frac{2}{2^*}} \le \left[\int_\Omega |T_k(u)|^{(2\lambda+1)m'}\right]^{\frac{1}{m'}} \|f\|_{L^m(\Omega)}\,.$$

Now it is sufficient to choose λ such that $(\lambda+1)2^* = m'(2\lambda+1)$, that is,

$$\lambda = \frac{-mN+2N-2m}{4m-2N}.$$

Using that $\frac{2}{2^*} - \frac{1}{m'} > 0$, one deduces that

$$\|T_k(u)\|_{L^{m^{**}}(\Omega)} \le C(\alpha, S, m, N)\,\|f\|_{L^m(\Omega)}\,.$$

Fatou's Lemma, as $k \to \infty$, implies the result. □

Let us now study the limit case where $f \in L^{\frac{N}{2}}(\Omega)$.

Theorem 6.10. *Let $f \in L^{\frac{N}{2}}(\Omega)$. Then every solution $u \in H_0^1(\Omega)$ to problem (6.3.1) is such that $e^{\lambda|u|}$ belongs $L^{2^*}(\Omega)$ for every $\lambda > 0$.*

Proof. Let us take $v = [e^{2\lambda|G_k(u)|} - 1]\mathrm{sgn}(G_k(u))$, $k > 0$ as a test function in the weak formulation of problem (6.3.1) and study the two sides separately. We can estimate the right-hand side using the following inequality, satisfied by every $t \ge 0$ and every $Q > 1$:

$$|t^2 - 1| \le Q(t-1)^2 + \frac{1}{Q-1}\,.$$

We then obtain

$$\int_\Omega f\,[e^{2\lambda|G_k(u)|} - 1]\mathrm{sgn}(G_k(u)) \le Q\int_{A_k} |f|[e^{\lambda|G_k(u)|} - 1]^2 + \frac{1}{Q-1}\int_{A_k} |f|\,.$$

Hölder's inequality with exponent $\frac{N}{2}$ implies

$$\int_\Omega f\,[e^{2\lambda|G_k(u)|} - 1]\mathrm{sgn}(G_k(u))$$

$$\le Q\|f\|_{L^{\frac{N}{2}}(A_k)}\left[\int_\Omega [e^{\lambda|G_k(u)|} - 1]^{2^*}\right]^{\frac{2}{2^*}} + \frac{1}{Q-1}\int_{A_k} |f|\,.$$

As a is elliptic, the left-hand side can be estimated from below by

$$2\lambda\alpha\int_\Omega |\nabla G_k(u)|^2 e^{2\lambda|G_k(u)|} = 2\lambda\alpha\int_\Omega \frac{1}{\lambda^2}\left|\nabla\left(e^{\lambda|G_k(u)|}-1\right)\right|^2.$$

Sobolev's inequality yields

$$\int_\Omega a(x,u,\nabla u)\cdot\nabla[e^{2\lambda|G_k(u)|}-1]\operatorname{sgn}(G_k(u)) \ge \frac{S^2}{\lambda}2\alpha\left[\int_\Omega [e^{\lambda|G_k(u)|}-1]^{2^*}\right]^{\frac{2}{2^*}}.$$

Therefore by the previous estimates we get

$$\begin{aligned}&\frac{S^2}{\lambda}2\alpha\left[\int_\Omega [e^{\lambda|G_k(u)|}-1]^{2^*}\right]^{\frac{2}{2^*}}\\ &\qquad\le Q\|f\|_{L^{\frac{N}{2}}(A_k)}\left[\int_\Omega [e^{\lambda|G_k(u)|}-1]^{2^*}\right]^{\frac{2}{2^*}} + \frac{1}{Q-1}\int_{A_k}|f|,\end{aligned}$$

that is,

$$\frac{S^2}{\lambda^2}\left(2\lambda\alpha - \frac{\lambda^2 Q}{S^2}\|f\|_{L^{\frac{N}{2}}(A_k)}\right)\left[\int_\Omega [e^{\lambda|G_k(u)|}-1]^{2^*}\right]^{\frac{2}{2^*}} \le \frac{1}{Q-1}\int_{A_k}|f|.$$

There exists k_λ such that

$$2\lambda\alpha - \frac{\lambda^2 Q\|f\|_{L^{\frac{N}{2}}(A_k)}}{S^2} > 0, \quad \forall\, k \ge k_\lambda$$

since $\|f\|_{L^{\frac{N}{2}}(A_k)} \to 0$ se $k \to +\infty$. Therefore the previous inequality implies that the sequence $e^{\lambda|G_k(u)|}-1$, for $k \ge k_\lambda$, is bounded in $L^{2^*}(\Omega)$. It is easy to deduce that $e^{\lambda|u|}$ belongs to $L^{2^*}(\Omega)$. Indeed

$$\begin{aligned}[e^{\lambda|u|}-1]^{2^*} &= [e^{\lambda|T_k(u)+G_k(u)|}-1]^{2^*} \le [e^{\lambda k}e^{\lambda|G_k(u)|}-e^{\lambda k}+e^{\lambda k}-1]^{2^*}\\ &\le 2^{2^*-1}e^{\lambda k 2^*}[e^{\lambda|G_k(u)|}-1]^{2^*} + 2^{2^*-1}[e^{\lambda k}-1]^{2^*}.\end{aligned}$$

Therefore, for every $k \ge k_\lambda$

$$\int_\Omega [e^{\lambda|u|}-1]^{2^*} \le 2^{2^*-1}e^{\lambda k 2^*}\int_\Omega [e^{\lambda|G_k(u)|}-1]^{2^*} + 2^{2^*-1}[e^{\lambda k}-1]^{2^*}\operatorname{meas}(\Omega),$$

that is, $e^{\lambda|u|}$ belongs to $L^{2^*}(\Omega)$ for every $\lambda > 0$. □

6.4 Sources in Marcinkiewicz spaces

We are now going to study the regularity of the weak solutions to (6.3.1), in the case where f belongs to the Marzinkiewicz space $M^m(\Omega)$.

Theorem 6.11. *Let $f \in M^m(\Omega)$, with $m > \frac{N}{2}$. Then any solution $u \in H_0^1(\Omega)$ to problem (6.3.1) is bounded.*

Proof. The result follows from Theorem 6.6, and from the inclusion $M^m(\Omega) \subset L^{m-\varepsilon}(\Omega)$ $(0 < \varepsilon < m-1)$, proved in Proposition 3.12. □

Theorem 6.12. *Let $f \in M^m(\Omega)$ with $(2^*)' < m < N/2$. Then any solution $u \in H_0^1(\Omega)$ to problem (6.3.1) belongs to $M^{m^{**}}(\Omega)$.*

Proof. We take $v = G_k(u)$ as a test function in the weak formulation of problem (6.3.1). We study separately the two sides of the equation. For the first one, we have

$$\int_\Omega a(x,u,\nabla u)\cdot\nabla G_k(u) \ge \alpha \int_{A_k} |\nabla u|^2$$
$$\ge \alpha S^2 \left(\int_\Omega |G_k(u)|^{2^*} \right)^{\frac{2}{2^*}}$$

by using the ellipticity of a and Sobolev's inequality. For the right-hand side, we get

$$\int_\Omega f\, G_k(u) \le \left(\int_\Omega |G_k(u)|^{2^*} \right)^{1/2^*} \left(\int_{A_k} |f|^{\frac{2N}{N+2}} \right)^{\frac{N+2}{2N}}$$
$$\le c \left(\int_\Omega |G_k(u)|^{2^*} \right)^{\frac{1}{2^*}} \operatorname{meas}(A_k)^{\left[1-\frac{2N}{(N+2)m}\right]\frac{N+2}{2N}},$$

where c is a constant depending on m, $\|f\|_{M^m(\Omega)}$), by using Hölder's inequality with exponent 2^* and Proposition 3.13 (applied to $|f|^{\frac{2N}{N+2}} \in M^{\frac{m(N+2)}{2N}}$). We have thus obtained

$$\left(\int_\Omega |G_k(u)|^{2^*} \right)^{\frac{1}{2^*}} \le \frac{c}{\alpha S^2} \operatorname{meas}(A_k)^{\frac{1}{2}+\frac{1}{N}-\frac{1}{m}}.$$

Using Hölder's inequality on the left-hand side of the above estimate we get

$$\int_\Omega |G_k(u)| \le \frac{c}{\alpha S^2} \operatorname{meas}(A_k)^{1+\frac{2}{N}-\frac{1}{m}}.$$

By Lemma 6.1, this is equivalent to

$$-\left(\frac{\alpha S^2}{c} \right)^{\nu} \ge g'(k) g(k)^{-\nu}$$

where

$$\nu = \frac{mN}{mN + 2m - N} > 1\,.$$

Integrating over $(0, k)$ we get

$$(\nu - 1)\left(\frac{\alpha\, S^2}{c}\right)^{\nu} k \ge [g(k)]^{1-\nu} - [g(0)]^{1-\nu}\,,$$

that is,

$$g(k) \le \left\{(\nu - 1)\left(\frac{\alpha\, S^2}{c}\right)^{\nu} k + [g(0)]^{1-\nu}\right\}^{\frac{1}{1-\nu}} \le \frac{C(\alpha, N, m, \Omega)}{k^{\frac{1}{\nu-1}}}\,,$$

since $\nu > 1$. We note that $A_{2k} \subset A_k$, and so

$$g(k) \ge \int_{A_{2k}} |G_k(u)|\, dx \ge \int_{A_{2k}} (|u| - k)\, dx \ge k\text{meas}(A_{2k})\,; \tag{6.4.1}$$

therefore

$$\text{meas}(A_{2k}) \le \frac{C(\alpha, N, m, \Omega)}{k^{\frac{\nu}{\nu-1}}}\,,$$

that is, $u \in M^{m^{**}}(\Omega)$. □

Theorem 6.13. *Let $f \in M^{\frac{N}{2}}(\Omega)$. Then there exists a constant $b > 0$ such that*

$$\int_{\Omega} e^{b\,|u|} < \infty$$

for every solution $u \in H_0^1(\Omega)$ to problem (6.3.1).

Proof. Choosing $v = G_k(u)$ as a test function in the weak formulation of (6.3.1), and arguing as in the proof of the previous theorem, we get

$$1 \le -\frac{c}{\alpha\, S^2}\frac{g'(k)}{g(k)}\,.$$

Integrating with respect to k we have

$$\int_0^k dt \le -\frac{c}{\alpha\, S^2}\int_0^k \frac{g'(t)}{g(t)}\, dt\,,$$

that is,

$$k \le -\frac{c}{\alpha\, S^2}[\ln g(k) - \ln g(0)] = \frac{c}{\alpha\, S^2}\ln\left(\frac{\|u\|_{L^1(\Omega)}}{g(k)}\right)\,.$$

This implies that

$$e^{k\alpha\, S^2/c} \le \frac{\|u\|_{L^1(\Omega)}}{g(k)}\,.$$

If $k \ge 1$, we get, arguing as in (6.4.1)

$$k\,\text{meas}(A_{2k}) \le g(k) \le \frac{\|u\|_{L^1(\Omega)}}{e^{k\alpha\, S^2/c}}\,.$$

The theorem follows from Lemma 6.5. □

6.5 Sources in divergence form

In this section, we focus our attention on

$$\begin{cases} -\mathrm{div}(a(x,u,\nabla u)) = -\mathrm{div} F\,, & \text{in} \quad \Omega\,, \\ u = 0\,, & \text{on} \quad \partial\Omega\,, \end{cases} \tag{6.5.1}$$

under the hypothesis that $F : \Omega \to \mathbb{R}^N$ is a vector field such that $\mathrm{div}F \in L^2(\Omega)$. We note that the existence of a weak solution in $H_0^1(\Omega)$ is guaranteed by the Leray–Lions theorem (Theorem 5.1).

Theorem 6.14. *Let $F \in (L^m(\Omega))^N, m > N$. Then any weak solution $u \in H_0^1(\Omega)$ to problem* (6.5.1) *is bounded; moreover, the estimate*

$$\|u\|_{L^\infty(\Omega)} \leq C\,\|F\|_{L^m(\Omega)}$$

holds, where C depends on N, m and α.

Proof. Take $G_k(u)$ as a test function in the weak formulation of (6.5.1):

$$\int_\Omega a(x,u,\nabla u)\cdot\nabla G_k(u) = \int_\Omega F\cdot\nabla G_k(u)\,.$$

Using the ellipticity of a on the left-hand side and the Cauchy–Schwarz inequality on the right one we get

$$\alpha\int_\Omega |\nabla G_k(u)|^2 \leq \left(\int_{A_k} |F|^2\right)^{1/2}\left(\int_\Omega |\nabla G_k(u)|^2\right)^{1/2},$$

where $A_k = \{|u| > k\}$. Using Hölder's inequality with exponent $m/2$ on the right-hand side, one has

$$\alpha\left[\int_\Omega |\nabla G_k(u)|^2\right]^{\frac{1}{2}} \leq \left(\int_{A_k} |F|^m\right)^{1/m} \mathrm{meas}(A_k)^{\frac{1}{2}-\frac{1}{m}}\,;$$

Sobolev's inequality allows us to say that

$$\alpha S\left(\int_\Omega |G_k(u)|^{2^*}\right)^{1/2^*} \leq ||F||_{L^m(\Omega)}\mathrm{meas}(A_k)^{\frac{1}{2}-\frac{1}{m}}\,. \tag{6.5.2}$$

We again use Hölder's inequality on the left-hand side of (6.5.2):

$$\begin{aligned} \int_\Omega |G_k(u)| &\leq \left(\int_\Omega |G_k(u)|^{2^*}\right)^{1/2^*} \mathrm{meas}(A_k)^{1-1/2^*} \\ &= \left(\int_\Omega |G_k(u)|^{2^*}\right)^{1/2^*} \mathrm{meas}(A_k)^{\frac{1}{2}+\frac{1}{N}}\,. \end{aligned}$$

Estimate (6.5.2) therefore yields

$$\int_\Omega |G_k(u)| \le C(\alpha, S)||F||_{L^m(\Omega)} \text{meas}(A_k)^{1-\frac{1}{m}+\frac{1}{N}}.$$

Using Lemma 6.2 we get the result. □

In the case $m < N$, we prove the following result:

Theorem 6.15. *Let $F \in (L^m(\Omega))^N$, $2 < m < N$. Then any weak solution $u \in H_0^1(\Omega)$ to problem* (6.5.1) *belongs to $L^{m^*}(\Omega)$; moreover the estimate*

$$\|u\|_{L^{m^*}(\Omega)} \le C\,\|F\|_{L^m(\Omega)}$$

holds, where C depends on N, m and α.

Proof. Let us consider $v = |T_k(u)|^{2\gamma}T_k(u)$ as a test function in the weak formulation of problem (6.5.1). We have, due to the ellipticity of a

$$\alpha\int_\Omega |\nabla T_k(u)|^2|T_k(u)|^{2\gamma} \le \int_\Omega F\cdot\nabla T_k(u)|T_k(u)|^{2\gamma}.$$

Hölder's inequality on the right-hand side yields

$$\alpha\int_\Omega |\nabla T_k(u)|^2|T_k(u)|^{2\gamma} \le \left(\int_\Omega |F|^2|T_k(u)|^{2\gamma}\right)^{1/2}\left(\int_\Omega |\nabla T_k(u)|^2|T_k(u)|^{2\gamma}\right)^{1/2},$$

and consequently

$$\alpha^2\int_\Omega |\nabla T_k(u)|^2|T_k(u)|^{2\gamma} \le \int_\Omega |F|^2|T_k(u)|^{2\gamma}.$$

Using again Hölder's inequality with exponent $m/2$ on the right-hand side, one gets

$$\alpha^2\left(\int_\Omega \left|\nabla|T_k(u)|^{\gamma+1}\right|^2\right) \le ||F||^2_{L^m(\Omega)}\left(\int_\Omega |T_k(u)|^{\frac{2m\gamma}{m-2}}\right)^{\frac{m-2}{m}}.$$

We use Sobolev's inequality on the left-hand side of the previous inequality:

$$\alpha^2S^2\left(\int_\Omega |T_k(u)|^{(\gamma+1)2^*}\right)^{2/2^*} \le ||F||^2_{L^m(\Omega)}\left(\int_\Omega |T_k(u)|^{\frac{2m\gamma}{m-2}}\right)^{\frac{m-2}{m}}. \tag{6.5.3}$$

Now, let us choose γ in such a way that

$$(\gamma+1)\,2^* = \frac{2m\gamma}{m-2}.$$

With this choice, estimate (6.5.3) can now be written as

$$\left(\int_\Omega |T_k(u)|^{m^*}\right)^{\frac{2}{2^*}-\frac{m-2}{m}} \le \frac{1}{\alpha^2 S^2}\,||F||^2_{L^m(\Omega)}\,.$$

We note that $\frac{2}{2^*} - \frac{m-2}{m} > 0$, since $m < N$. Fatou's lemma implies

$$||u||_{L^{m^*}(\Omega)} \le C\,||F||_{L^m(\Omega)},$$

with $0 < C = C(\alpha, N, m)$. □

7 H^2 regularity for linear problems

7.1 Introduction

In this chapter, we focus our attention on the regularity of the solutions to linear elliptic problems. We study

$$\begin{cases} -\operatorname{div}(M(x)\nabla u) = f\,, & \text{in} \quad \Omega\,, \\ u = 0\,, & \text{on} \quad \partial\Omega\,, \end{cases} \tag{7.1.1}$$

where Ω is an open bounded subset of $\mathbb{R}^N, N \geq 3$, and $f \in L^2(\Omega)$. Moreover $M(x)$ is a $\mathbb{R}^{N\times N}$ matrix such that $M(x)\xi \cdot \xi \geq \alpha|\xi|^2$ for every $\xi \in \mathbb{R}^N$ and the entries m_{ij} of M are Lipschitz continuous, that is,

$$|m_{ij}(x) - m_{ij}(y)| \leq K\,|x-y|\,, \quad \forall\, x, y \in \overline{\Omega}\,, \quad \forall\, i, j = 1, ..N\,.$$

In Chapter 4, we proved the existence of a unique distributional solution $u \in H_0^1(\Omega)$. Now we prove that u belongs to H^2, following the proof by Nirenberg, given in [40].

Theorem 7.1. *Let $u \in H_0^1(\Omega)$ be the solution to problem* (7.1.1). *Then, for every $\Omega' \subset\subset \Omega$, u belongs to $H^2(\Omega')$ and the following estimate holds:*

$$\|u\|_{H^2(\Omega')} \leq C\,(\|u\|_{H_0^1(\Omega)} + \|f\|_{L^2(\Omega)}) \tag{7.1.2}$$

where $C = C(\Omega, K, \alpha, d)$, with $d = dist(\Omega', \partial\Omega)$.

7.2 Preliminaries

We define the incremental rate of a function. We denote by $e_i \in \mathbb{R}^N$ the vector having all the coefficients 0, except the ith, which is 1.

Definition 7.2. Let $f : \Omega \to \mathbb{R}$ and $h \in \mathbb{R} \setminus \{0\}$. The incremental rate of f with respect to e_i is the function

$$\Delta_i^h f : \{x \in \Omega : x + he_i \in \Omega\} \to \mathbb{R}\,,$$

defined by

$$\Delta_i^h f(x) := \frac{f(x + he_i) - f(x)}{h}\,.$$

We remark that $\Delta_i^h f$ is defined in

$$\Omega_{|h|} := \{x \in \Omega : \operatorname{dist}(x, \partial\Omega) > |h|\}\,.$$

From now on, we will write Δ^h instead of Δ_i^h. We will denote by $\nabla_i f$ the ith component of the vector ∇f.

Proposition 7.3. *The incremental rate of a function has the following properties:*
(1) *If $f \in W^{1,p}(\Omega)$, then $\Delta^h f \in W^{1,p}(\Omega_{|h|})$ and*

$$\nabla_i(\Delta^h f) = \Delta^h(\nabla_i f)\,. \tag{7.2.1}$$

(2) *If one of the functions f or g has support contained in $\Omega_{|h|}$, then*

$$\int_\Omega f\Delta_i^h g = -\int_\Omega g\Delta_i^{-h} f\,. \tag{7.2.2}$$

(3) *The following equality holds:*

$$\Delta_i^h(fg)(x) = f(x+he_i)\Delta_i^h g(x) + g(x)\Delta_i^h f(x)\,. \tag{7.2.3}$$

Lemma 7.4. *Let $v \in W^{1,p}(\Omega)$. Then $\Delta^h v \in L^p(\Omega')$ for every $\Omega' \subset\subset \Omega$ such that $h < dist(\Omega', \partial\Omega)$ and the following estimate holds:*

$$\|\Delta_i^h v\|_{L^p(\Omega')} \le \|\nabla_i v\|_{L^p(\Omega')}\,.$$

Proof. Assume that $v \in C^1(\Omega) \cap W^{1,p}(\Omega)$. Then

$$\Delta_i^h v(x) = \frac{v(x+he_i) - v(x)}{h} = \frac{1}{h}\int_0^h \nabla_i v(x_1,\dots,x_{i-1},x_i+\xi,x_{i+1},\dots,x_N)d\xi\,.$$

Hölder's inequality implies

$$|\Delta_i^h v(x)|^p \le \frac{1}{h}\int_0^h |\nabla_i v(x_1,\dots,x_{i-1},x_i+\xi,x_{i+1},\dots,x_N|^p d\xi$$

and so

$$\int_{\Omega'} |\Delta_i^h v|^p \le \frac{1}{h}\int_0^h\int_{\Omega'} |\nabla_i v|^p\, d\xi \le \int_{\Omega'} |\nabla_i v|^p\,.$$

Using the density of $C^1(\Omega) \cap W^{1,p}(\Omega)$ in $W^{1,p}(\Omega)$ we can prove the general result. □

We now prove that the previous estimate is sufficient for f to be in $W_0^{1,p}(\Omega)$, in some sense.

Lemma 7.5. *Let $v \in L^p(\Omega)$, $1 < p < \infty$, and suppose that there exists K such that $\|\Delta^h v\|_{L^p(\Omega')} \le K$ for every $h > 0$ and $\Omega' \subset\subset \Omega$ such that $h < dist(\Omega', \partial\Omega)$. Then the distributional gradient ∇v exists and satisfies $\|\nabla v\|_{L^p(\Omega)} \le K$.*

Proof. Let $h_n \to 0$; let us define, for $i = 1, \ldots, N$,

$$g_n := \begin{cases} \Delta_i^{h_n} v & \text{in} \quad \Omega_{|h_n|} \\ 0 & \text{in} \quad \Omega \setminus \Omega_{|h_n|} \,. \end{cases}$$

The sequence g_n is bounded in $L^p(\Omega)$ and so there exists a subsequence, still denoted by g_n, which converges weakly to $\tilde{v} \in L^p(\Omega)$, with $\|\tilde{v}_i\|_{L^p(\Omega)} \le K$. Therefore for every $\varphi \in C_0^1(\Omega)$ one has

$$\int_\Omega \varphi \Delta_i^{h_n} v \to \int_\Omega \varphi \tilde{v}_i \,.$$

On the other hand, for $h_n < \operatorname{dist}(\operatorname{supp}\varphi, \partial\Omega)$, we have, using (7.2.2) and Lebesgue's theorem

$$\int_\Omega \varphi \Delta_i^{h_n} v = -\int_\Omega v \Delta_i^{-h_n} \varphi \to -\int_\Omega v \nabla_i \varphi, \quad n \to \infty \,.$$

We deduce that for every $\varphi \in C_0^1(\Omega)$

$$\int_\Omega \varphi \tilde{v}_i = -\int_\Omega v \nabla_i \varphi$$

and so $\tilde{v}_i = \nabla_i v$. □

7.3 $H^2(\Omega)$ regularity of the solutions

This section is devoted to the proof of Theorem 7.1. Note that this result allows us to say that u solves $-\operatorname{div}(M(x)\nabla u) = f$ a.e. in $\Omega' \subset\subset \Omega$, since $u \in H^2(\Omega')$.

Proof of Theorem 7.1. The solution u satisfies

$$\sum_{i,j=1}^N \int_\Omega m_{ij}(x) \nabla_j u \nabla_i v = \int_\Omega f v \,, \quad \forall\, v \in H_0^1(\Omega) \,.$$

Let φ be a function with compact support in Ω and $|2h| < \operatorname{dist}(\operatorname{supp}\varphi, \partial\Omega)$. Let $v = \Delta_k^{-h}\varphi$, for $1 \le k \le N$: properties (7.2.1) and (7.2.2) imply

$$\sum_{i,j=1}^N \int_\Omega \Delta_k^h (m_{ij} \nabla_j u) \nabla_i \varphi = -\sum_{i,j=1}^N \int_\Omega m_{ij} \nabla_j u \nabla_i \Delta_k^{-h} \varphi = -\int_\Omega f \Delta_k^{-h} \varphi \,.$$

Using property (7.2.3),

$$\Delta_k^h (m_{ij} \nabla_j u)(x) = m_{ij}(x + h e_k) \Delta_k^h \nabla_j u(x) + \nabla_j u(x) \Delta_k^h m_{ij}(x)$$

one has

$$\sum_{i,j=1}^{N}\int_{\Omega} m_{ij}(x+he_k)\Delta_k^h\nabla_j u(x)\nabla_i\varphi$$

$$= -\sum_{i,j=1}^{N}\int_{\Omega}\left[\Delta_k^h m_{ij}(x)\nabla_j u(x)\nabla_i\varphi + m_{ij}(x)\nabla_j u(x)\nabla_i\Delta_k^{-h}\varphi\right]$$

$$= -\sum_{i,j=1}^{N}\int_{\Omega}[\Delta_k^h m_{ij}(x)\nabla_j u(x)\nabla_i\varphi + f\Delta_k^{-h}\varphi]\,.$$

Therefore

$$\sum_{i,j=1}^{N}\int_{\Omega} m_{ij}(x+he_k)\Delta_k^h\nabla_j u(x)\nabla_i\varphi = -\int_{\Omega}[g\cdot\nabla\varphi + f\Delta_k^{-h}\varphi]\,,$$

where $g = (g^1,\dots,g^n)$ with $g^i = \Delta_k^h m_{ij}\nabla_j u$. Using the Cauchy–Schwarz inequality and Lemma 7.4 we get

$$\sum_{i,j=1}^{N}\int_{\Omega} m_{ij}(x+he_k)\Delta_k^h\nabla_j u(x)\nabla_i\varphi \le (K\|\nabla u\|_{L^2(\Omega)} + \|f\|_{L^2(\Omega)})\|\nabla\varphi\|_{L^2(\Omega)}\,,$$

since m_{ij} are Lipschitz continuous functions. Property (7.2.1) implies

$$\sum_{i,j=1}^{N}\int_{\Omega} m_{ij}(x+he_k)\nabla_j\Delta_k^h u(x)\nabla_i\varphi \le (K\|u\|_{H_0^1(\Omega)} + \|f\|_{L^2(\Omega)})\|\nabla\varphi\|_{L^2(\Omega)}\,. \quad (7.3.1)$$

Let $\eta \in C_0^1(\Omega)$ be such that $0 \le \eta \le 1$ and define $\varphi = \eta^2\Delta_k^h u$. Writing $\nabla_i\varphi$ in an explicit way and using (7.2.1), we get

$$\sum_{i,j=1}^{N}\int_{\Omega}\eta^2 m_{ij}(x+he_k)\nabla_i\Delta_k^h u\nabla_j\Delta_k^h u$$

$$= \sum_{i,j=1}^{N}\int_{\Omega} m_{ij}(x+he_k)\nabla_j\Delta_k^h u(\nabla_i\varphi - 2\Delta_k^h u\eta\nabla_i\eta)\,. \quad (7.3.2)$$

The ellipticity of M gives

$$\alpha\int_{\Omega}|\eta\nabla\Delta_k^h u|^2 \le \sum_{i,j=1}^{N}\int_{\Omega}\eta^2 m_{i,j}(x+he_k)\Delta_k^h\nabla_i u\Delta_k^h\nabla_j u\,.$$

As for the right-hand side of (7.3.2) one has, due to (7.3.1) and the Cauchy–Schwarz inequality

$$\begin{aligned}
&\sum_{i,j=1}^{N}\int_\Omega m_{i,j}(x+he_k)\nabla_j\Delta_k^h u(\nabla_i\varphi-2\Delta_k^h u\eta\nabla_i\eta)\\
&\leq \sum_{i,j=1}^{N}\int_\Omega m_{i,j}(x+he_k)\nabla_j\Delta_k^h u\nabla_i\varphi+2\left|\int_\Omega m_{i,j}(x+he_k)\nabla_j\Delta_k^h u\Delta_k^h u\eta\nabla_i\eta\right|\\
&\leq (K\|u\|_{H_0^1(\Omega)}+\|f\|_{L^2(\Omega)})\|\nabla(\eta^2\Delta_k^h u)\|_{L^2(\Omega)}+2K\|\eta\nabla\Delta_k^h u\|_{L^2(\Omega)}\|\Delta_k^h u\nabla\eta\|_{L^2(\Omega)}\\
&\leq (K\|u\|_{H_0^1(\Omega)}+\|f\|_{L^2(\Omega)})\|2\eta\nabla\eta\Delta_k^h u+\eta^2\nabla\Delta_k^h u\|_{L^2(\Omega)}\\
&\quad+2K\|\eta\nabla\Delta_k^h u\|_{L^2(\Omega)}\|\Delta_k^h u\nabla\eta\|_{L^2(\Omega)}\,.
\end{aligned}\tag{7.3.3}$$

We remark that

$$\|2\eta\nabla\eta\Delta_k^h u+\eta^2\nabla\Delta_k^h u\|_{L^2(\Omega)}\leq\|2\nabla\eta\Delta_k^h u\|_{L^2(\Omega)}+\|\eta\nabla\Delta_k^h u\|_{L^2(\Omega)}$$

since $0\leq\eta\leq 1$. We deduce from the previous estimates and (7.3.3) that

$$\begin{aligned}
&\sum_{i,j}^{N}\int_\Omega m_{i,j}(x+he_k)\nabla_j\Delta_k^h u(\nabla_i\varphi-2\Delta_k^h u\eta\nabla_i\eta)\\
&\quad\leq (K\|u\|_{H_0^1(\Omega)}+\|f\|_{L^2(\Omega)})(\|2\nabla\eta\Delta_k^h u\|_{L^2(\Omega)}+\|\eta\nabla\Delta_k^h u\|_{L^2(\Omega)})\\
&\qquad+2K\|\eta\nabla\Delta_k^h u\|_{L^2(\Omega)}\|\Delta_k^h u\nabla\eta\|_{L^2(\Omega)}\,.
\end{aligned}$$

Therefore we have got from (7.3.2)

$$\begin{aligned}
\alpha\|\eta\nabla\Delta_k^h u\|_{L^2(\Omega)}^2\leq{}&\|\eta\nabla\Delta_k^h u\|_{L^2(\Omega)}(K\|u\|_{H_0^1(\Omega)}+\|f\|_{L^2(\Omega)})\\
&+2\|\Delta_k^h u\nabla\eta\|_{L^2(\Omega)}(K\|u\|_{H_0^1(\Omega)}\\
&+\|f\|_{L^2(\Omega)})+2K\|\eta\nabla\Delta_k^h u\|_{L^2(\Omega)}\|\Delta_k^h u\nabla\eta\|_{L^2(\Omega)}\,.
\end{aligned}$$

It follows from the Young inequality that

$$\begin{aligned}
\alpha\|\eta\nabla\Delta_k^h u\|_{L^2(\Omega)}^2\leq{}&\frac{1}{2\varepsilon}(K\|u\|_{H_0^1(\Omega)}+\|f\|_{L^2(\Omega)})^2+\frac{\varepsilon}{2}\|\eta\nabla\Delta_k^h u\|_{L^2(\Omega)}^2\\
&+\frac{1}{\varepsilon}(K\|u\|_{H_0^1(\Omega)}+\|f\|_{L^2(\Omega)})^2+\varepsilon\|\Delta_k^h u\nabla\eta\|_{L^2(\Omega)}^2\\
&+\varepsilon K\|\eta\nabla\Delta_k^h u\|_{L^2(\Omega)}^2+\frac{1}{\varepsilon}\|\Delta_k^h u\nabla\eta\|_{L^2(\Omega)}^2\,,
\end{aligned}$$

or equivalently,

$$\begin{aligned}
&\left(\alpha-\frac{\varepsilon}{2}-\varepsilon K\right)\|\eta\nabla\Delta_k^h u\|_{L^2(\Omega)}^2\\
&\qquad\leq\frac{3}{2\varepsilon}(K\|u\|_{H_0^1(\Omega)}+\|f\|_{L^2(\Omega)})^2+\left(\varepsilon+\frac{1}{\varepsilon}\right)\|\Delta_k^h u\nabla\eta\|_{L^2(\Omega)}^2\,.
\end{aligned}$$

Choosing ε sufficiently small, one gets, if c denotes a constant depending just on k and α

$$\begin{aligned}\|\eta\nabla\Delta_k^h u\|_{L^2(\Omega)}^2 &\leq c\,(K\|u\|_{H_0^1(\Omega)} + \|f\|_{L^2(\Omega)})^2 + c\|\Delta_k^h u\nabla\eta\|_{L^2(\Omega)}^2\\ &\leq c\,(\|u\|_{H_0^1(\Omega)} + \|f\|_{L^2(\Omega)} + \|\Delta_k^h u\nabla\eta\|_{L^2(\Omega)})^2.\end{aligned}$$

By using (7.2.1)

$$\|\eta\Delta_k^h\nabla u\|_{L^2(\Omega)} \leq c\,(\|u\|_{H_0^1(\Omega)} + \|f\|_{L^2(\Omega)} + \sup_{\Omega}|\nabla\eta|\|\Delta_k^h u\|_{L^2(spt\,\eta)})$$

where $spt\,\eta$ is the support of η. Lemma 7.4 implies

$$\|\eta\Delta_k^h\nabla u\|_{L^2(\Omega)} \leq c\,(\|u\|_{H_0^1(\Omega)} + \|f\|_{L^2(\Omega)} + \sup_{\Omega}|\nabla\eta|\|\nabla u\|_{L^2(\Omega)})$$

and so, for every $\Omega' \subset\subset \Omega$ fixed

$$\|\eta\Delta_k^h\nabla u\|_{L^2(\Omega')} \leq c\,(1 + \sup_{\Omega}|\nabla\eta|)(\|u\|_{H_0^1(\Omega)} + \|f\|_{L^2(\Omega)})\,.$$

Now, η can be chosen as a cut-off function, such that $\eta = 1$ on Ω' and $|\nabla\eta| \leq \frac{2}{d}$, where $d = \mathrm{dist}(\Omega', \partial\Omega)$. From Lemma 7.5 we get that $\nabla u \in H^1(\Omega')$ for every $\Omega' \subset\subset \Omega$, and so $u \in H^2(\Omega)$. Estimate (7.1.2) is thus proved, using Poincaré's inequality. □

8 Spectral analysis for linear operators

8.1 Introduction

In this chapter, we will focus on the eigenvalue problem

$$\begin{cases} -\mathrm{div}(M(x)\nabla u) = \lambda u\,, & \text{in } \Omega\,, \\ u = 0\,, & \text{on } \partial\Omega\,, \end{cases} \tag{8.1.1}$$

under the following assumptions. Ω is an open bounded subset of $\mathbb{R}^N, N \geq 3$, and $M(x)$ is a $N \times N$ symmetric matrix, with bounded entries such that there exists $\alpha > 0$ satisfying

$$M(x)\xi \cdot \xi \geq \alpha|\xi|^2$$

for every $\xi \in \mathbb{R}^N$. We will first study the existence and some properties of the eigenvalues of $L(v) = -\mathrm{div}(M(x)\nabla v)$. Later we will see some applications of the spectral theory to some semilinear (noncoercive) equations. We will present some existence and multiplicity results by Dolph [36], Ambrosetti and Prodi [2].

8.2 Eigenvalues of linear elliptic operators

Theorem 8.1. *There exists an orthonormal basis $w_m \in L^2(\Omega)$ and a sequence of real positive numbers λ_m such that*
(1) $\lambda_m \to +\infty$, *as $m \to +\infty$;*
(2) *for every $m \in \mathbb{N}$, w_m is a solution to*

$$\begin{cases} -\mathrm{div}(M(x)\nabla v) = \lambda_m v\,, & \text{in } \Omega\,, \\ v = 0\,, & \text{on } \partial\Omega\,. \end{cases}$$

The proof of this theorem is based on important properties of

$$\begin{aligned} T:\ \ L^2(\Omega) &\to L^2(\Omega) \\ f &\to u \end{aligned} \tag{8.2.1}$$

where $u \in H_0^1(\Omega)$ solves $-\mathrm{div}(M(x)\nabla u) = f$.

Lemma 8.2. *Let T be defined by* (8.2.1). *Then T is self-adjoint and compact.*

Proof. T is well defined by Theorem 4.4 and is linear. We claim that T is self-adjoint (according to Definition 8.22). Let $U = T(u)$ and $V = T(v)$. Then for every $\varphi \in H_0^1(\Omega)$

$$\int_\Omega M(x)\nabla U \cdot \nabla\varphi = \int_\Omega u\,\varphi$$

and

$$\int_\Omega M(x)\nabla V \cdot \nabla\varphi = \int_\Omega v\,\varphi .$$

Let us choose $\varphi = V$ in the first equation and $\varphi = U$ in the second one. The symmetry of M implies that

$$\int_\Omega u\,V = \int_\Omega v\,U ,$$

that is, T is self-adjoint. Let us prove that T is compact. If $T(f) = u$, then

$$\alpha\|\nabla u\|^2_{L^2(\Omega)} \le \int_\Omega M(x)\nabla u\cdot\nabla u = \int_\Omega fu \le \|f\|_{L^2(\Omega)}\|u\|_{L^2(\Omega)} \le c\,\|f\|_{L^2(\Omega)}\|\nabla u\|_{L^2(\Omega)} ,$$

using the ellipticity of M on the left-hand side and Hölder's and Poincaré's inequality on the right one. We deduce that

$$\|T(f)\|_{H_0^1(\Omega)} \le \frac{c}{\alpha}\|f\|_{L^2(\Omega)} , \quad \forall\, f \in L^2(\Omega) .$$

Since the embedding $H_0^1(\Omega) \hookrightarrow L^2(\Omega)$ is compact, T is compact. □

We can now prove Theorem 8.1.

Proof. Using Lemma 8.2, we can apply the spectral theorem (Theorem 8.25) to the operator T defined by (8.2.1). Therefore there exists an orthogonal basis w_n of $L^2(\Omega)$ with

$$\int_\Omega |w_n|^2 = 1 \tag{8.2.2}$$

and a sequence μ_n converging to 0, as $n \to \infty$, such that $T(w_n) = \mu_n w_n$, that is,

$$\int_\Omega M(x)\nabla w_n \cdot \nabla\varphi = \frac{1}{\mu_n}\int_\Omega w_n\varphi , \quad \forall\,\varphi \in H_0^1(\Omega) .$$

Observe that $w_n \in H_0^1(\Omega)$. Moreover $\mu_n \ge 0$, since by the ellipticity of M, we have

$$\alpha\int_\Omega |\nabla w_n|^2 \le \frac{1}{\mu_n}\int_\Omega w_n^2 .$$

On the other hand $\mu_n \ne 0$, otherwise $w_n = 0$ from $T(w_n) = 0$. □

Definition 8.3. According to the notations of the previous theorem, we will say that $\{\lambda_m\}_{m\in\mathbb{N}}$ is the set of the eigenvalues of $L(v) = -\mathrm{div}(M(x)\nabla v)$: this means that $\{\frac{1}{\lambda_m}\}_{m\in\mathbb{N}}$ is the set of eigenvalues of

$$\begin{aligned} T:\ & L^2(\Omega) \to L^2(\Omega) \\ & f \to u \end{aligned}$$

where $u \in H_0^1(\Omega)$ satisfies $-\mathrm{div}(M(x)\nabla u) = f$. By an eigenfunction of $L(v) = -\mathrm{div}(M(x)\nabla v)$ we shall mean an eigenfunction of T.

The following theorem concerns λ_1, the first eigenvalue of $L(v) = -\mathrm{div}(M(x)\nabla v)$. Let

$$A(v) = \frac{\int_\Omega M(x)\nabla v \cdot \nabla v}{\int_\Omega v^2}.$$

Theorem 8.4. *Let λ_1 be the smallest eigenvalue of $L(v) = -div(M(x)\nabla v)$. Then*

$$\lambda_1 = \min_{\substack{v\in H_0^1(\Omega)\\ v\neq 0}} A(v).$$

Moreover, each function u that minimizes A is an eigenfunction of L corresponding to λ_1.

Remark 8.5. Poincaré's inequality, for $p = 2$ reads: there exists a positive constant $c = c(\Omega)$ such that

$$\frac{1}{c}\|u\|_{L^2(\Omega)} \leq \|\nabla u\|_{L^2(\Omega)}, \quad \forall u \in H_0^1(\Omega).$$

Theorem 8.4 allows us to say that the smallest eigenvalue for the operator $L(v) = -\Delta v$ is equal to the square root of the best constant in Poincaré's inequality.

We will use the following lemma to prove Theorem 8.4:

Lemma 8.6. *$A(v)$ has a minimizer over $H_0^1(\Omega)$.*

Proof. First of all we observe that A is bounded from below, by Poincaré's inequality. Let v_n be a minimizing sequence, that is, $A(v_n) \to \inf A$, as $n \to +\infty$. Note that

$$z_n = \frac{v_n}{\|v_n\|_{H_0^1(\Omega)}}$$

satisfies $\|z_n\|_{H_0^1(\Omega)} = 1$. Up to a subsequence, $z_n \to z$ in $L^2(\Omega)$ and weakly in $H_0^1(\Omega)$. Observe that

$$0 \leq M(x)\nabla(z_n - z)\cdot\nabla(z_n - z) = M(x)\nabla z_n\cdot\nabla z_n - 2M(x)\nabla z_n\cdot\nabla z + M(x)\nabla z\cdot\nabla z.$$

Therefore,

$$\int_\Omega M(x)\nabla z_n \cdot \nabla z_n \geq 2\int_\Omega M(x)\nabla z_n \cdot \nabla z - \int_\Omega M(x)\nabla z \cdot \nabla z.$$

If we pass to the limit in the above inequality, we have

$$\liminf_{n\to\infty}\int_\Omega M(x)\nabla z_n \cdot \nabla z_n \geq \int_\Omega M(x)\nabla z \cdot \nabla z. \tag{8.2.3}$$

Note that z_n is a minimizing sequence for A, as A is homogeneous. Thus

$$\inf A = \lim_{n\to\infty} A(z_n) = \liminf_{n\to\infty}\frac{\int_\Omega M(x)\nabla z_n \cdot \nabla z_n}{\int_\Omega z_n^2} \geq \frac{\int_\Omega M(x)\nabla z \cdot \nabla z}{\int_\Omega z^2} = A(z),$$

where we have used (8.2.3) and the fact that $z_n \to z$ in $L^2(\Omega)$. Consequently z is a minimizer for A. We have only to check that $z \neq 0$. Since z_n is a minimizing sequence, $A(z_n)$ is a bounded sequence, that is, there exists a positive constant C such that

$$\int_\Omega M(x)\nabla z_n \cdot \nabla z_n \leq C \int_\Omega z_n^2 .$$

The ellipticity of M gives

$$\alpha \int_\Omega |\nabla z_n|^2 \leq C \int_\Omega z_n^2 .$$

This implies that

$$1 = \|z_n\|^2_{L^2(\Omega)} + \|\nabla z_n\|^2_{L^2(\Omega)} \leq \left(1 + \frac{C}{\alpha}\right) \|z_n\|^2_{L^2(\Omega)} .$$

At the limit as $n \to \infty$ we get $\|z\|_{L^2(\Omega)} > 0$. □

We can now prove Theorem 8.4.

Proof. Let $u \in H_0^1(\Omega)$ be a minimizer for A. Using the previous lemma the function $g(t) = A(u + tw)$, where $w \in H_0^1(\Omega)$ attains its minimum at 0. Since g is differentiable, $g'(0) = 0$, that is,

$$\int_\Omega M(x)\nabla u \cdot \nabla w = \frac{\int_\Omega M(x)\nabla u \cdot \nabla u}{\int_\Omega u^2} \int_\Omega uw = (\inf A) \int_\Omega uw , \quad \forall\, w \in H_0^1(\Omega) . \tag{8.2.4}$$

In other words, if $u \in H_0^1(\Omega)$ minimizes A, then u is an eigenfunction of $L(v) = -\mathrm{div}(M(x)\nabla v)$ and

$$\frac{\int_\Omega M(x)\nabla u \cdot \nabla u}{\int_\Omega u^2}$$

is the corresponding eigenvalue.

Let us prove that $\inf A = \lambda_1$. Since λ_1 is the smallest eigenvalue of $-\mathrm{div}(M(x)\nabla v)$, one has

$$\lambda_1 \leq \inf A .$$

Let us prove the opposite inequality. Let w_1 be the eigenfunction corresponding to λ_1; we have

$$\int_\Omega M(x)\nabla w_1 \cdot \nabla v = \lambda_1 \int_\Omega w_1 v , \quad \forall\, v \in H_0^1(\Omega) .$$

Consequently, choosing $v = w_1$ we obtain

$$\inf A = \inf \frac{\int_\Omega M(x)\nabla v \cdot \nabla v}{\int_\Omega v^2} \leq \frac{\int_\Omega M(x)\nabla w_1 \cdot \nabla w_1}{\int_\Omega w_1^2} = \frac{\int_\Omega \lambda_1 w_1^2}{\int_\Omega w_1^2} = \lambda_1 .$$ □

Remark 8.7. As we will see in Chapter 9, Example 9.9, (8.2.4) is the Euler equation for the functional A.

Corollary 8.8. *Any eigenfunction w_1 of $L(v) = -div(M(x)\nabla v)$ corresponding to λ_1 is of constant sign on Ω.*

Proof. By the definition of eigenfunction corresponding to λ_1, w_1 solves

$$\int_\Omega M(x)\nabla w_1 \cdot \nabla v = \lambda_1 \int_\Omega w_1 v\,, \quad \forall\, v \in H_0^1(\Omega)\,.$$

Considering $v = w_1^+$ one has

$$\lambda_1 = \frac{\int_\Omega M(x)\nabla w_1^+ \cdot \nabla w_1^+}{\int_\Omega (w_1^+)^2}\,,$$

that is, w_1^+ is a minimum for A. By Theorem 8.4, w_1^+ is an eigenfunction of $L(v) = -\text{div}(M(x)\nabla v)$ corresponding to λ_1. □

Remark 8.9. One can prove more about λ_1 (see [37]). Indeed λ_1 is simple and the corresponding eigenfunctions are strictly positive (negative). This is a consequence of Harnack's inequality (see [50]): if u is a nonnegative solution to $L(u) = -\text{div}(M(x)\nabla u) = \lambda u$, then for any compact $G \subset \Omega$ there exists a positive constant K, independent of u, such that $\max_G u \le K \min_G u$. The constant K depends on α, the $L^\infty(\Omega)$ norm of the matrix M, λ, G, and Ω.

More generally, we have the following expression for λ_m:

Theorem 8.10. *Let $m > 1$. Then*

$$\lambda_m = \frac{\int_\Omega M(x)\nabla w_m \cdot \nabla w_m}{\int_\Omega w_m^2} = \inf_{v\in P_{m-1}, v\neq 0} \frac{\int_\Omega M(x)\nabla v \cdot \nabla v}{\int_\Omega v^2}$$

where $P_m = \left\{v \in H_0^1(\Omega) : \int_\Omega w_n\, v = 0\,, \quad n = 1, \dots, m\right\}$.

We can now prove the last result of this section: the eigenfunctions of (8.1.1) are bounded.

Theorem 8.11. *Let u be an eigenfunction of $L(v) = -div(M(x)\nabla v)$ corresponding to an eigenvalue λ. Then u is bounded and the following estimate holds:*

$$\|u\|_{L^\infty(\Omega)} \le c(\alpha, N)\lambda^{\frac{N}{2}} \|u\|_{L^1(\Omega)}\,. \tag{8.2.5}$$

Proof. We take $G_k(u) = u - T_k(u)$ as a test function in (8.1.1): in this way

$$\alpha \int_{A_k} |\nabla G_k(u)|^2 \le \int_\Omega M(x)\nabla u \cdot \nabla u = \lambda \int_\Omega u\, G_k(u)\,,$$

by the ellipticity of M. Let us now estimate the right-hand side. Writing u as $u-k+k$, we get, by Young's inequality

$$\begin{aligned}\lambda\int_\Omega uG_k(u) &\le \lambda\int_{A_k}|G_k(u)|^2+\lambda\int_{A_k}k\,|G_k(u)|\\ &\le \lambda\int_{A_k}|G_k(u)|^2+\frac{\lambda}{2}\left\{\int_{A_k}|G_k(u)|^2+k^2\text{meas}(A_k)\right\}\\ &=3\frac{\lambda}{2}\int_{A_k}|G_k(u)|^2+k^2\frac{\lambda}{2}\text{meas}(A_k)\,,\end{aligned}$$

where $A_k=\{|u|\ge k\}$. We have thus proved that

$$\alpha\int_{A_k}|\nabla G_k(u)|^2\le 3\frac{\lambda}{2}\int_{A_k}|G_k(u)|^2+k^2\,\frac{\lambda}{2}\,\text{meas}(A_k)\,. \tag{8.2.6}$$

We are going to estimate the first term of the right-hand side by using Hölder's and Sobolev's inequalities:

$$\begin{aligned}S\left[\int_{A_k}|G_k(u)|^2\right]^{\frac12} &\le S\left[\int_{A_k}|G_k(u)|^{2^*}\right]^{\frac{1}{2^*}}\text{meas}(A_k)^{\frac1N}\\ &\le\left[\int_{A_k}|\nabla G_k(u)|^2\right]^{\frac12}\text{meas}(A_k)^{\frac1N}\,.\end{aligned} \tag{8.2.7}$$

Inequality (8.2.6) implies that

$$\alpha\int_{A_k}|\nabla G_k(u)|^2\le\frac{3\lambda}{2S^2}\text{meas}(A_k)^{\frac2N}\int_{A_k}|\nabla G_k(u)|^2+k^2\frac{\lambda}{2}\text{meas}(A_k)\,.$$

Let us now consider $k\ge k_0$, where $k_0=k_0(\lambda)$ is such that

$$\alpha\ge\frac{3\lambda}{S^2}\text{meas}(A_{k_0})^{\frac2N}\,.$$

Note that $k_0\,\text{meas}(A_{k_0})\le\|u\|_{L^1(\Omega)}$. Therefore, in the sequel we will consider

$$k\ge k_0=\left[\frac{3\lambda}{S^2\alpha}\right]^{\frac N2}\|u\|_{L^1(\Omega)}\,. \tag{8.2.8}$$

In this way

$$\alpha\int_{A_k}|\nabla G_k(u)|^2\le k^2\,\lambda\,\text{meas}(A_k)\,. \tag{8.2.9}$$

From estimates (8.2.9) and (8.2.7) we deduce that

$$\left[\int_{A_k} |G_k(u)|^2\right]^{\frac{1}{2}} \le \frac{k}{S}\left[\frac{\lambda \operatorname{meas}(A_k)}{\alpha}\right]^{\frac{1}{2}} \operatorname{meas}(A_k)^{\frac{1}{N}}.$$

By Hölder's inequality on the left-hand side one has

$$\int_{A_k} |G_k(u)| \le \left[\int_{A_k} |G_k(u)|^2\right]^{\frac{1}{2}} \operatorname{meas}(A_k)^{\frac{1}{2}} \le \operatorname{meas}(A_k)^{1+\frac{1}{N}} k (S^2\alpha)^{-\frac{1}{2}}\lambda^{\frac{1}{2}}.$$

Setting $g(k) = \int_{A_k} |G_k(u)|$ and using Lemma 6.1, the last estimate is equivalent to

$$S\,\alpha^{\frac{1}{2}}\, g(k) \le [-g'(k)]^{1+\frac{1}{N}}\, k\,\lambda^{\frac{1}{2}},$$

that is,

$$g'(k)\, g(k)^{-\frac{N}{N+1}}\, \lambda^{\frac{N}{2(N+1)}} \le -k^{-\frac{N}{N+1}}\, (S\alpha^{\frac{1}{2}})^{\frac{N}{N+1}}.$$

Integrating over (k_0, k) we have

$$g(k)^{\frac{1}{N+1}} \le g(k_0)^{\frac{1}{N+1}} + (S\,\alpha^{\frac{1}{2}})^{\frac{N}{N+1}}\,\lambda^{-\frac{N}{2(N+1)}}\,[k_0^{\frac{1}{N+1}} - k^{\frac{1}{N+1}}].$$

Since $g(k_0) \le \|u\|_{L^1(\Omega)}$ we get

$$g(k)^{\frac{1}{N+1}} \le \|u\|_{L^1(\Omega)}^{\frac{1}{N+1}} + (S\,\alpha^{\frac{1}{2}})^{\frac{N}{N+1}}\,\lambda^{-\frac{N}{2(N+1)}}\,[k_0^{\frac{1}{N+1}} - k^{\frac{1}{N+1}}].$$

For $k = \tilde{k}$,

$$\tilde{k} = \left[(S\,\alpha^{\frac{1}{2}})^{-\frac{N}{N+1}}\,\lambda^{\frac{N}{2(N+1)}}\,\|u\|_{L^1(\Omega)}^{\frac{1}{N+1}} + k_0^{\frac{1}{N+1}}\right]^{N+1}, \tag{8.2.10}$$

the right-hand side is zero. Thus $g(k) = 0$, if $k \ge \tilde{k}$. Observe that $\tilde{k} \ge k_0$.

Note that, by using (8.2.8), we get

$$\tilde{k} \le c(\alpha, N)\,\lambda^{\frac{N}{2}}\,\|u\|_{L^1(\Omega)}.$$

Therefore

$$\|u\|_{L^\infty(\Omega)} \le c(\alpha, N)\,\lambda^{\frac{N}{2}}\,\|u\|_{L^1(\Omega)}. \qquad \square$$

Remark 8.12. That the eigenfunctions of $L(v) = -\operatorname{div}(M(x)\nabla v)$ are bounded is a consequence of the following bootstrapping technique. By definition, $w_m \in H_0^1$ is a solution to

$$\begin{cases} -\operatorname{div}(M(x)\nabla w_m) = \lambda_m w_m, & \text{in } \Omega, \\ w_m = 0, & \text{on } \partial\Omega. \end{cases}$$

Since $H_0^1(\Omega) \subseteq L^{2^*}(\Omega)$, using Theorem 6.9 we have $w_m \in L^{2^{***}}(\Omega)$. Using Theorem 6.9 in an iterative way we get, after a finite number of steps, the $L^p(\Omega)$ summability with $p > N/2$, for any $N \geq 3$. To this end, we define

$$\begin{cases} q_0 = 2^* \\ \quad \vdots \\ q_{k+1} = q_k^{**} = \frac{Nq_k}{N-2q_k}, \quad k \geq 0. \end{cases}$$

By contradiction, assume that

$$q_k \leq N/2$$

for every k. Since q_k is strictly monotone, there exists

$$l := \lim q_k .$$

One has, necessarily

$$0 < l \leq N/2 .$$

Passing to the limit on k, we get

$$l = \frac{Nl}{N-2l} :$$

this equality implies $l = 0$ which is a contradiction. Therefore there exists $\bar{k} \geq 0$ such that $q_{\bar{k}} > N/2$. It is sufficient to use Theorem 6.6 to get the result. We note that the previous theorem gives us an additional information with respect to the above bootstrapping technique: estimate (8.2.5).

Corollary 8.13. *Let u be an eigenfunction of $L(v) = -div(M(x)\nabla v)$ corresponding to an eigenvalue λ. Then*

$$\|u\|_{L^\infty(\Omega)} \leq c(\alpha, N)\,\lambda^{\frac{N}{2}}\,\mathrm{meas}(\Omega)^{\frac{1}{2}} .$$

Proof. It is sufficient to apply Hölder's inequality on the right-hand side of estimate (8.2.5) and to use (8.2.2). □

8.3 Applications to some semilinear equations

In this section we will study some semilinear (noncoercive) equations using the results of spectral analysis of the previous section. More precisely we will study

$$\begin{cases} -\mathrm{div}(M(x)\nabla u) = g(u) + f, & \text{in } \Omega, \\ u = 0, & \text{on } \partial\Omega, \end{cases} \tag{8.3.1}$$

under different hypotheses on g. Note that we have already studied some semilinear equations in Chapter 4. We will begin by the linear problem

$$\begin{cases} -\text{div}(M(x)\nabla u) = \mu u + f, & \text{in } \Omega, \\ u = 0, & \text{on } \partial\Omega, \end{cases} \tag{8.3.2}$$

where $\mu \in \mathbb{R}$.

Theorem 8.14. *Let $\mu \in \mathbb{R}$ with $\mu \neq \lambda_k$ for every $k \in \mathbb{N}$, where $\{\lambda_k\}_{k\in\mathbb{N}}$ are the eigenvalues of $L(v) = -div(M(x)\nabla v)$. Then for every $f \in L^2(\Omega)$ there exists a unique solution w to problem* (8.3.2).

Proof. We can suppose that $\mu \neq 0$, because we have already seen in Theorem 4.4 that there exists a unique $H_0^1(\Omega)$ solution to problem $-\text{div}(M(x)\nabla u) = f$. Let T be the operator defined by (8.2.1). Then μ^{-1} is not an eigenvalue of T. Consequently, using Theorem 8.27 (Fredholm alternative), for every $f \in L^2(\Omega)$ there exists a unique solution to the equation

$$\mu^{-1}u - Tu = \mu^{-1}Tf,$$

that is, there exists a unique solution to problem (8.3.2). □

Theorem 8.15. *Let $\mu = \lambda_k$, where for some $k \in \mathbb{N}$, λ_k is an eigenvalue of $L(v) = -div(M(x)\nabla v))$. Let $f \in L^2(\Omega)$ be such that $\int_\Omega f\, w_k = 0$ for every eigenfunction w_k corresponding to λ_k. Then there exists a solution to problem* (8.3.2).

Proof. As in the above theorem we will use the operator T defined by (8.2.1). By Theorem 8.27 (Fredholm alternative), there exists a solution to

$$\mu^{-1}u - Tu = \mu^{-1}Tf$$

(i. e., to problem (8.3.2)) provided $\int_\Omega Tf\varphi = 0$ for every φ such that $\mu^{-1}\varphi = T\varphi$, that is, for every eigenfunction of $L(v) = -\text{div}(M(x)\nabla v))$ corresponding to μ. Now, since T is self-adjoint, one has

$$0 = \int_\Omega Tf\varphi = \int_\Omega f\,T\varphi = \mu^{-1}\int_\Omega f\varphi.$$ □

The existence of the eigenfunctions of $L(v) = -\text{div}(M(x)\nabla v)$ is sometimes useful to find a subsolution (or a supersolution) in view of an application of Theorem 4.11, as the following theorem shows.

Theorem 8.16. *Let $\theta \in (0,1)$. Then there exists a positive solution $u \in H_0^1(\Omega)$ to problem*

$$\begin{cases} -\Delta u = u^\theta, & \textit{in } \Omega, \\ u = 0, & \textit{on } \partial\Omega. \end{cases} \tag{8.3.3}$$

Proof. Let $\psi \in H_0^1(\Omega)$ be the positive solution to problem

$$\begin{cases} -\Delta\psi = 1\,, & \text{in} \quad \Omega\,, \\ \psi = 0\,, & \text{on} \quad \partial\Omega\,. \end{cases}$$

From Theorem 6.6 we know that ψ is bounded; then for any $a > 0$ such that

$$a^{\frac{1-\theta}{\theta}} \geq \|\psi\|_{L^\infty(\Omega)}\,,$$

$\overline{u} = a\psi$ is a supersolution to problem (8.3.3). On the other hand, let φ_1 be an eigenfunction corresponding to the first eigenvalue λ_1 of $L(v) = -\Delta v$: φ_1 can be chosen positive by Theorem 8.8 and is bounded by Theorem 8.11. Choosing $t > 0$ such that

$$\lambda_1(t\,\varphi_1)^{1-\theta} \leq 1\,,$$

one has that $\underline{u} = t\,\varphi_1$ is a subsolution to problem (8.3.3). To prove that $\underline{u} \leq \overline{u}$, we remark that, by linearity,

$$-\Delta(a\,\psi - t\,\varphi_1) = a - \lambda_1 t\,\varphi_1\,.$$

It is then sufficient to consider T such that $T \geq \lambda_1 t\|\varphi_1\|_{L^\infty(\Omega)}$: in this way $\overline{u} \geq \underline{u}$ from Lemma 4.12. The result follows from Theorem 4.11. □

In the following two theorems, the hypotheses on the function g in problem (8.3.1) are related to the eigenvalues of $L(v) = -\text{div}(M(x)\nabla v)$.

We are now going to state Dolph's theorem.

Theorem 8.17 (Dolph). *Let $g : \mathbb{R} \to \mathbb{R}$ be a function with the property that there exists $\delta > 0$ such that*

$$0 < \lambda_k + \delta < \frac{g(t) - g(s)}{t - s} < \lambda_{k+1} - \delta\,,$$

for some $k \in \mathbb{N}$, where $\{\lambda_k\}$ are the eigenvalues of $L(v) = -div(M(x)\nabla v)$. Then, for every $f \in L^2(\Omega)$, there exists a solution $u \in H_0^1(\Omega)$ to problem (8.3.1).

In the proof we will use the following result:

Lemma 8.18. *Let $\mu \in \mathbb{R}$ with $\mu \neq \lambda_k$. Let S be defined by*

$$\begin{aligned} S:\ \ L^2(\Omega) &\to L^2(\Omega) \\ f &\to w \end{aligned}$$

where $w \in H_0^1(\Omega)$ solves $-div(M(x)\nabla w) = \mu w + f$. Then S is compact.

Remark 8.19. Observe that S is well defined by Theorem 8.14.

Proof. It is sufficient to prove that $S : L^2(\Omega) \to H_0^1(\Omega)$ is continuous, as the embedding $H_0^1(\Omega) \hookrightarrow L^2(\Omega)$ is compact. We will prove, by linearity, that if $f_n \to 0$ in $L^2(\Omega)$, then the corresponding solutions z_n to problem (8.3.2) converge to 0 in $H_0^1(\Omega)$. Let us consider a test function in (8.3.2) φ_i, the eigenfunctions of $L(v) = -\text{div}(M(x)\nabla v)$: one has

$$(\lambda_i - \mu)\int_\Omega z_n\varphi_i = \int_\Omega M(x)\nabla z_n \cdot \nabla\varphi_i - \mu\int_\Omega z_n\,\varphi_i = \int_\Omega f_n\varphi_i\,.$$

Since $f_n \to 0$ in $L^2(\Omega)$ we have $\int_\Omega z_n\varphi_i \to 0$, as $n \to \infty$. The sequence $\{\varphi_i\}$ is an orthonormal basis of $L^2(\Omega)$, as proved in Theorem 8.1: this implies that $z_n \to 0$ weakly in $L^2(\Omega)$. Consequently, choosing z_n as test function in (8.3.2), one has

$$\alpha\int_\Omega |\nabla z_n|^2 \le \int_\Omega M(x)\nabla z_n \cdot \nabla z_n = \mu\int_\Omega z_n^2 + \int_\Omega f_n z_n\,. \tag{8.3.4}$$

Since the right-hand side is uniformly bounded, up to a subsequence, z_n has a weak limit in $H_0^1(\Omega)$ which is necessarily 0 and $z_n \to 0$ in $L^2(\Omega)$. Consequently the right-hand side of (8.3.4) goes to 0 and so $z_n \to 0$ in $H_0^1(\Omega)$. This proves the lemma. □

We now prove Dolph's theorem:

Proof. Problem (8.3.1) is equivalent to

$$\begin{cases} -\text{div}(M(x)\nabla u) - \lambda u = g(u) - \lambda u + f\,, & \text{in } \Omega\,,\\ u = 0\,, & \text{on } \partial\Omega\,,\end{cases}$$

where $\lambda = \frac{\lambda_{k+1}+\lambda_k}{2}$. We observe that $\lambda \neq 0$ and that λ is not an eigenvalue of L. This means that $\frac{1}{\lambda}$ is not an eigenvalue of the operator T defined by (8.2.1). Let

$$\begin{aligned} S:\ & L^2(\Omega) \to L^2(\Omega)\\ & f \to w \end{aligned}$$

where w solves $-\text{div}(M(x)\nabla w) - \lambda w = f$. By Lemma 8.18 S is continuous.

Setting

$$\begin{aligned} \Theta:\ & L^2(\Omega) \to L^2(\Omega)\\ & v \to S[g(v) - \lambda v + f] \end{aligned}$$

it is sufficient to prove that Θ has a fixed point. We are going to apply Theorem 2.1. To do that we are going to prove that S is a contraction. The linearity and the continuity of S imply

$$\|\Theta v - \Theta w\|_{L^2(\Omega)} \le \|S\|_{\mathcal{L}(L^2(\Omega))}\,\|g(v) - g(w) - \lambda(v - w)\|_{L^2(\Omega)}\,. \tag{8.3.5}$$

Let us now estimate $\|g(v) - g(w) - \lambda(v - w)\|_{L^2(\Omega)}$. From the hypotheses on g we deduce

$$\left|\frac{g(v) - g(w)}{v - w} - \lambda\right| \le \frac{\lambda_{k+1} - \lambda_k}{2} - \delta\ :$$

this is equivalent to

$$|g(v) - g(w) - \lambda(v - w)| \leq \left(\frac{\lambda_{k+1} - \lambda_k}{2} - \delta\right) |v - w|,$$

and so

$$\|g(v) - g(w) - \lambda(v - w)\|_{L^2(\Omega)} \leq \left(\frac{\lambda_{k+1} - \lambda_k}{2} - \delta\right) \|v - w\|_{L^2(\Omega)}.$$

By (8.3.5), one has

$$\|\Theta v - \Theta w\|_{L^2(\Omega)} \leq \|S\|_{\mathcal{L}(L^2(\Omega))} \left(\frac{\lambda_{k+1} - \lambda_k}{2} - \delta\right) \|v - w\|_{L^2(\Omega)}. \tag{8.3.6}$$

We claim that

$$\|S\|_{\mathcal{L}(L^2(\Omega))} \left(\frac{\lambda_{k+1} - \lambda_k}{2} - \delta\right) < 1. \tag{8.3.7}$$

Let us estimate $\|S\|_{\mathcal{L}(L^2(\Omega))}$. We denote by ν_k the eigenvalues of $-\text{div}(M(x)\nabla v) - \lambda v$, that is,

$$-\text{div}(M(x)\nabla z_k) - \lambda z_k = \nu_k z_k.$$

This equality implies that $\nu_k = \lambda_k - \lambda$, where λ_k are the eigenvalues of $L(v) = -\text{div}(M(x)\nabla v)$. The eigenvalues of S are therefore

$$\nu_k^{-1} = \frac{1}{\lambda_k - \lambda}.$$

Theorem 8.26 implies that

$$\|S\|_{\mathcal{L}(L^2(\Omega))} = \sup_i \left|\frac{1}{\lambda_i - \lambda}\right|$$

for S is compact, as seen in Lemma 8.18. Since $0 < \cdots < \lambda_k < \lambda < \lambda_{k+1} < \ldots$ we have

$$\|S\|_{\mathcal{L}(L^2(\Omega))} = \sup_{i=k,k+1} \left|\frac{1}{\lambda_i - \lambda}\right|.$$

Moreover the fact that $\lambda - \lambda_k = \lambda_{k+1} - \lambda$ implies that

$$\|S\|_{\mathcal{L}(L^2(\Omega))} = \frac{1}{\lambda_{k+1} - \lambda} = \frac{2}{\lambda_{k+1} - \lambda_k}.$$

We then deduce (8.3.7). Hence Θ is a contraction. Theorem 2.1 implies that there exists a unique $u \in L^2(\Omega)$ such that

$$-\text{div}(M(x)\nabla u) + \lambda u = g(u) + \lambda u + f.$$

Since $S : L^2(\Omega) \to H_0^1(\Omega)$ we have that $u \in H_0^1(\Omega)$, that is, u is the solution. □

We are now going to prove the following theorem:

Theorem 8.20 (Ambrosetti–Prodi). *Let $g : \mathbb{R} \to \mathbb{R}$ be a Lipschitz continuous function such that $g(0) = 0$. Assume that*

$$\lim_{s \to \pm\infty} \frac{g(s)}{s} = \gamma_\pm, \quad \gamma_- < \lambda_1 < \gamma_+ < \lambda_2,$$

where λ_1 and λ_2 are the first two eigenvalues of $L(v) = -\operatorname{div}(M(x)\nabla v)$. Then for every $f \in L^2(\Omega)$, there exists $\bar{t} \in \mathbb{R}$ such that:
(1) *if $\int_\Omega f\varphi_1 > \bar{t}$, there is no solution to problem* (8.3.1)*;*
(2) *if $\int_\Omega f\varphi_1 = \bar{t}$, there exists a solution to problem* (8.3.1)*;*
(3) *if $\int_\Omega f\varphi_1 < \bar{t}$, there exist two solutions to problem* (8.3.1).

We will use the following result.

Lemma 8.21. *Under the same hypotheses on g as in Theorem 8.20, let $u_n, u \in L^2(\Omega)$ be such that $u_n \to u$ in $L^2(\Omega)$, as $n \to +\infty$.*
(1) *If $t_n \to +\infty$, then $\frac{g(t_n u_n)}{t_n} \to \gamma_+ u^+ - \gamma_- u^-$ in $L^2(\Omega)$.*
(2) *If $t_n \to -\infty$, then $\frac{g(t_n u_n)}{t_n} \to \gamma_- u^+ - \gamma_+ u^-$ in $L^2(\Omega)$.*

Proof. Set

$$\rho_n(x) = \begin{cases} \frac{g(t_n u_n(x))}{t_n u_n(x)}, & \text{if} \quad u_n(x) \neq 0, \\ 0, & \text{if} \quad u_n(x) = 0. \end{cases}$$

We observe that the sequence ρ_n is bounded, due to the hypotheses on g. Consequently $\rho_n(u_n - u) \to 0$ in $L^2(\Omega)$.
(1) Assume that $t_n \to +\infty$. Studying separately

$$\Omega_n^- = \{x \in \Omega : u_n(x) < 0\}$$
$$\Omega_n^0 = \{x \in \Omega : u_n(x) = 0\}$$
$$\Omega_n^+ = \{x \in \Omega : u_n(x) > 0\}$$

one gets that $\rho_n u \to \gamma_+ u^+ - \gamma_- u_-$ a.e. in Ω. Lebesgue's theorem implies that $\rho_n u \to \gamma_+ u^+ - \gamma_- u^-$ in $L^2(\Omega)$. Therefore $\rho_n u_n \to \gamma_+ u^+ - \gamma_- u^-$ in $L^2(\Omega)$.
(2) The case where $t_n \to -\infty$ is similar to the previous case. □

We can now prove Theorem 8.20.

Proof. Let φ_1 be a positive eigenfunction of $L(v) = -\operatorname{div}(M(x)\nabla v)$ corresponding to the first eigenvalue λ_1, such that $\|\varphi_1\|_{L^2(\Omega)} = 1$. We will prove that for every $s \in \mathbb{R}$ there exists a unique solution $z = z_s \in H_0^1(\Omega)$ to

$$\int_\Omega M(x)\nabla z \cdot \nabla w = \int_\Omega g(z + s\varphi_1)w + \int_\Omega fw, \quad \forall w \in H_0^1(\Omega) : \int_\Omega w\varphi_1 = 0 \quad (8.3.8)$$

and $\int_\Omega z\varphi_1 = 0$. We will afterward study the existence of a real s such that

$$\lambda_1 s - \int_\Omega g(z + s\,\varphi_1)\varphi_1 = \int_\Omega f\varphi_1 . \tag{8.3.9}$$

It is easily seen that the existence of a solution to problem (8.3.1) is equivalent to the existence of z and s.

Step I: Let us study problem (8.3.8). For every fixed $s \in \mathbb{R}$, we will find a unique solution z_s to problem (8.3.8) using Theorem 4.3. Indeed, define

$$a(\psi, w) = \int_\Omega M(x)\nabla\psi \cdot \nabla w - \int_\Omega g(\psi + s\,\varphi_1)w$$

on the Hilbert space made up of functions $w \in H_0^1(\Omega)$ such that $\int_\Omega w\varphi_1 = 0$. This form is linear in the second variable. Using the Cauchy–Schwarz inequality and the fact that g is Lipschitz continuous, we can find a positive constant C such that

$$|a(\psi_1, w) - a(\psi_2, w)| \le C\,\|\nabla w\|_{L^2(\Omega)}\,\|\nabla(\psi_1 - \psi_2)\|_{L^2(\Omega)} .$$

Moreover it is easy to see that

$$\begin{aligned} a(\psi_1, \psi_1 - \psi_2) &- a(\psi_2, \psi_1 - \psi_2) \\ &\ge \int_\Omega M(x)\nabla(\psi_1 - \psi_2)\cdot\nabla(\psi_1 - \psi_2) - \gamma_+ \int_\Omega |\psi_1 - \psi_2|^2 . \end{aligned}$$

By Theorem 8.10 and the ellipticity of M we deduce that

$$a(\psi_1, \psi_1 - \psi_2) - a(\psi_2, \psi_1 - \psi_2) \ge \left(1 - \frac{\gamma_+}{\lambda_2}\right)\alpha\|\nabla(\psi_1 - \psi_2)\|^2_{L^2(\Omega)} .$$

We can then use Theorem 4.3 and state that for every $s \in \mathbb{R}$ there exists a unique solution z_s to problem (8.3.8).

Step II: Let us prove that

$$h(s) = \lambda_1 s - \int_\Omega g(z_s + s\varphi_1)\varphi_1$$

is a continuous function. Obviously it is sufficient to prove that the second term is a continuous function from $\mathbb{R}$ to $\mathbb{R}$. To this end, let s_n be a real sequence. We claim that the corresponding sequence z_{s_n} of solutions to problem (8.3.8) is uniformly bounded in $H_0^1(\Omega)$. We can write

$$\int_\Omega M(x)\nabla z_{s_n}\cdot\nabla z_{s_n} = \int_\Omega g(z_{s_n} + s_n\varphi_1)z_{s_n} + \int_\Omega f\,z_{s_n} .$$

Multiplying and dividing by $z_{s_n} + s_n\varphi_1$ in the right-hand side and using the limit hypothesis on g, we have

$$\int_\Omega M(x)\nabla z_{s_n}\cdot\nabla z_{s_n} \le \gamma_+\int_\Omega z_{s_n}^2 + \|f\|_{L^2(\Omega)}\|z_{s_n}\|_{L^2(\Omega)}\,.$$

By Theorem 8.10 and the ellipticity of M we have

$$\alpha\left(1-\frac{\gamma_+}{\lambda_2}\right)\|\nabla z_{s_n}\|^2_{L^2(\Omega)} \le \|f\|_{L^2(\Omega)}\|z_{s_n}\|_{L^2(\Omega)}\,.$$

Poincaré's inequality implies that the sequence z_{s_n} is bounded in $H_0^1(\Omega)$. Now, assume that $s_n \to s_0$ as $n\to\infty$. Up to a subsequence, z_{s_n} converges weakly to a function w in $H_0^1(\Omega)$. Theorem 3.6 implies that

$$g(z_{s_n}+s_n\varphi_1)\to g(w+s_0\varphi_1)$$

in $L^2(\Omega)$. We have to prove that $w = z_{s_0}$, that is, w is the solution to problem (8.3.8) corresponding to s_0. Passing to the limit in

$$\int_\Omega M(x)\nabla z_{s_n}\cdot\nabla\psi = \int_\Omega g(z_{s_n}+s_n\varphi_1)\psi + \int_\Omega f\psi\,,\quad \forall\psi:\int_\Omega \psi\varphi_1 = 0$$

we get

$$\int_\Omega M(x)\nabla w\cdot\nabla\psi = \int_\Omega g(w+s_0\varphi_1)\psi + \int_\Omega f\psi\,.$$

From Step *I* there exists a unique solution to problem $-\mathrm{div}(M(x)\nabla z_{s_0}) = g(z_{s_0} + s_0\varphi_1) + f$ and so necessarily $w = z_{s_0}$. Consequently, up to a subsequence,

$$\int_\Omega g(z_{s_n}+s_n\varphi_1)\varphi_1 \to \int_\Omega g(z_{s_0}+s_0\varphi_1)\varphi_1\,.$$

Arguing by contradiction, it is easily seen that

$$\int_\Omega g(z_{s_n}+s_n\varphi_1)\varphi_1 \to \int_\Omega g(z_{s_0}+s_0\varphi_1)\varphi_1$$

and not only a subsequence. This implies that h is a continuous function.

Step III: Let us prove that

$$\lim_{s\to\pm\infty}\frac{h(s)}{s} = \lambda_1 - \gamma_\pm\,.$$

Since

$$\frac{h(s)}{s} = \lambda_1 - \int_\Omega \frac{g\left[s\left(\frac{z_s}{s}+\varphi_1\right)\right]}{s}\varphi_1\,,$$

it is sufficient to study the last term. We remark that z_s is uniformly bounded in $H_0^1(\Omega)$ as we have already proved in the previous step. Consequently $\frac{z_s}{s} \to 0$ in $H_0^1(\Omega)$ as $s \to \infty$. Setting $v_s = \frac{z_s}{s} + \varphi_1$ we have that $v_s \to \varphi_1$ in $L^2(\Omega)$.

(1) Assume that $s \to +\infty$. By Lemma 8.21, we have

$$\frac{g(sv_s)}{s} \to \gamma_+ \varphi_1^+ - \gamma_- \varphi_1^-$$

in $L^2(\Omega)$. Since φ_1 is positive,

$$\int_\Omega \frac{g\left[s\left(\frac{z_s}{s} + \varphi_1\right)\right]}{s}\varphi_1 \to \gamma_+ .$$

(2) In a similar way, one can study the case $s \to -\infty$.

Step IV: Problem (8.3.9) is equivalent to $h(s) = \int_\Omega f\varphi_1$. Since

$$\lambda_1 - \gamma_+ < 0 < \lambda_1 - \gamma_- ,$$

using the result of the previous step, we can say that h has a maximum. Therefore

(1) if $\int_\Omega f\varphi_1 < \max_{\mathbb{R}} h$, then problem (8.3.1) has at least two solutions;
(2) if $\int_\Omega f\varphi_1 = \max_{\mathbb{R}} h$, then problem (8.3.1) has at least one solution;
(3) if $\int_\Omega f\varphi_1 > \max_{\mathbb{R}} h$, then problem (8.3.1) has no solution. □

8.4 Appendix

We recall here some classical results of spectral theory for linear operators. For the proofs, see [22].

Definition 8.22. Let H be a Hilbert space. Let $T : H \to H$ be a linear operator. Let $B_H = \{x \in H : \|x\| \le 1\}$. We will say that

(1) T is self-adjoint if $(Tu|v) = (u|Tv)$ for every $u, v \in H$;
(2) T is compact if $T(B_H)$ is relatively compact for the strong topology.

Definition 8.23. Let E be a Banach space. Let $T : E \to E$ be a linear operator.

(1) we set $\rho(T) = \{\lambda \in \mathbb{R} : T - \lambda I : E \to E \quad \text{is one-to-one}\}$;
(2) the spectrum of T is $\sigma(T) = \mathbb{R} \setminus \rho(T)$;
(3) λ is an eigenvalue if $Ker(T - \lambda I) \neq 0$. We will denote by $AT(T)$ the eigenvalues set.

Remark that the eigenvalues set of T is contained in $\sigma(T)$, but it is not equal in general.

Theorem 8.24. *Let E be Banach space. Let $T : E \to E$ be a linear compact operator. Then*

(1) $0 \in \sigma(T)$;
(2) $\sigma(T) \setminus \{0\} = AT(T) \setminus \{0\}$;
(3) *either $\sigma(T) \setminus \{0\}$ is finite or $\sigma(T) \setminus \{0\}$ is a sequence which goes to 0.*

Theorem 8.25 (Spectral Theorem). *Let H be a Hilbert separable space. Let T be a linear, self-adjoint compact operator. Then H has an orthonormal basis composed on eigenvectors of T. Moreover, the sequence of the corresponding eigenvalues λ_n is such that $|\lambda_n| \to 0$ as $n \to \infty$.*

Theorem 8.26. *Let H be a Hilbert space. Let $T : H \to H$ be a linear, compact, self-adjoint operator. Then, if μ_n are the eigenvalues of T, one has*

$$\|T\|_{\mathcal{L}(H)} = \sup_n |\mu_n| .$$

Theorem 8.27 (Fredholm alternative). *Let H be a Hilbert space. Let $T : H \to H$ be a linear, compact, self-adjoint operator. Let $\lambda \in \mathbb{R} \setminus \{0\}$. Then the following alternative holds: either for any $\xi \in H$ the equation $\lambda x - Tx = \xi$ has a unique solution or else $\lambda x = Tx$ has solutions $x \neq 0$ and $\lambda x - Tx = \xi$ has a solution if $\xi \perp Ker(\lambda I - T)$.*

9 Calculus of variations and Euler's equation

9.1 Introduction

Although this book is devoted to partial differential equations, we would like to recall some results of the Calculus of Variations. We will see how the Calculus of Variations can be used to study the existence of solutions to differential problems.

We will first present a simple but significant version of De Giorgi's theorem on weak lower semicontinuity of integral functionals. We will then write their Euler's equation and study the summability of minimizers. In the last part of this chapter we will study Ekeland's principle.

9.2 Direct methods in the calculus of variations

In this section we will recall some classical results of the Calculus of Variations which will be useful to study the minimization of a functional. We refer to [27] for the proofs.

We will often use the following two theorems (in the Appendix we will recall some definitions).

Theorem 9.1 (Weierstrass). *Let X be a Banach reflexive space. Let $J : X \to \mathbb{R}$ be coercive, bounded from below and weakly lower semicontinuous. Then J has a minimizer.*

Proof. Let x_n be a minimizing sequence, that is, $J(x_n) \to \inf J$. Note that $J(x_n)$ is bounded. Since J is coercive, x_n is bounded in X. Therefore, up to a subsequence, $x_n \to x_0$ weakly in X, for some $x_0 \in X$, since X is reflexive. Since J is weakly lower semicontinuous, $\liminf_{n\to\infty} J(x_n) \geq J(x_0)$. Hence $J(x_0) = \inf J$. □

We will essentially consider integral functionalsin this chapter:

$$J(v) = \int_\Omega j(x, v, \nabla v)\,,$$

defined on $W_0^{1,p}(\Omega)$ with $p \in (1, \infty)$, with Ω bounded open subset of $\mathbb{R}^N$. Note that these spaces are reflexive. The following theorem gives a sufficient condition for the functional to be weakly lower semicontinuous in $W_0^{1,p}(\Omega)$ (see [26]).

Theorem 9.2 (De Giorgi). *Let $j : \Omega \times \mathbb{R} \times \mathbb{R}^N \to \mathbb{R}$ be a Carathéodory function, convex with respect to the last variable. Let $p > 1$ and $1 \leq q < p^*$ if $p < N$ and $1 \leq q < +\infty$ if $p \geq N$. Assume that there exists $\alpha_1 \in (L^{p'}(\Omega))^N$, $\alpha_2 \in \mathbb{R}$, $\alpha_3 \in L^1(\Omega)$ such that*

$$j(x, s, \xi) \geq \alpha_1(x) \cdot \xi + \alpha_2 |s|^q + \alpha_3(x)\,.$$

Let $u_n, u \in W_0^{1,p}(\Omega)$ be such that $u_n \to u$ weakly in $W^{1,p}(\Omega)$. Then

$$\int_\Omega j(x, u, \nabla u) \leq \liminf_{n\to+\infty} \int_\Omega j(x, u_n, \nabla u_n)\,.$$

The proof of this theorem is quite complicated. We will present it for a particular class of functionals. The symbol $\nabla_\xi j$ will denote the gradient of j with respect to ξ.

Theorem 9.3. *Let $j : \Omega \times \mathbb{R} \times \mathbb{R}^N \to \mathbb{R}$ be a Carathéodory, convex function with respect to the last variable. Assume that there exist $\alpha, \beta > 0$ such that $\alpha|\xi|^p \leq j(x,s,\xi) \leq \beta|\xi|^p$. Assume that for a.e. $x \in \Omega$ and for every $s \in \mathbb{R}$ $j(x,s,\cdot)$ is differentiable and there exists $\nu > 0$ such that $|\nabla_\xi j(x,s,\xi)| \leq \nu|\xi|^{p-1}$. Then $J(v) = \int_\Omega j(x,v,\nabla v)$, defined on $W_0^{1,p}(\Omega)$, has a minimizer.*

Proof. J is coercive and bounded from below, since $j(x,s,\xi) \geq \alpha|\xi|^p$. Let us prove that it is weakly lower semicontinuous. Let v_n be a sequence weakly converging to v in $W_0^{1,p}(\Omega)$. The hypotheses of convexity and differentiability on j imply

$$j(x,s,\xi) \geq j(x,s,\eta) + \nabla_\xi j(x,s,\eta) \cdot (\xi - \eta),$$

and therefore

$$\int_\Omega j(x,v_n,\nabla v_n) \geq \int_\Omega j(x,v_n,\nabla v) + \int_\Omega \nabla_\xi j(x,v_n,\nabla v) \cdot (\nabla v_n - \nabla v). \qquad (9.2.1)$$

By the continuity of $j(x,\cdot,\xi)$ and the growth conditions on j, we can apply Theorem 3.6 to have that $j(x,v_n,\nabla v)$ converges in $L^1(\Omega)$ to $j(x,v,\nabla v)$. Moreover $\nabla v_n - \nabla v$ converges weakly to 0 in $(L^p(\Omega))^N$ by hypothesis. By the continuity of $j(x,\cdot,\xi)$ and the growth conditions on $\nabla_\xi j$ we can apply again Theorem 3.6 to have that $\nabla_\xi j(x,v_n,\nabla v)$ converges to $\nabla_\xi j(x,v,\nabla v)$ in $(L^{p'}(\Omega))^N$. Passing to the lim inf in (9.2.1), we get

$$\liminf_{n\to+\infty} \int_\Omega j(x,v_n,\nabla v_n) \geq \int_\Omega j(x,v,\nabla v). \qquad \square$$

Example 9.4. Let J be defined on $H_0^1(\Omega)$ by

$$J(v) = \frac{1}{2}\int_\Omega |\nabla v|^2 - \frac{1}{p}\int_\Omega |v|^p - \int_\Omega fv$$

with $p \in [1,2)$ and $f \in L^2(\Omega)$. We claim that this functional has a minimizer. Note that, by Hölder's and Poincaré's inequalities

$$J(v) \geq \frac{1}{2}\|\nabla v\|^2_{L^2(\Omega)} - \frac{1}{p}\int_\Omega |v|^p - c\,\|f\|_{L^2(\Omega)}\|\nabla v\|_{L^2(\Omega)}.$$

By Rellich–Kondrachov theorem, there exists a positive constant C such that $\|v\|^p_{L^p(\Omega)} \leq C\|\nabla v\|^p_{L^2(\Omega)}$. Therefore

$$J(v) \geq \frac{1}{2}\|\nabla v\|^2_{L^2(\Omega)} - \frac{C}{p}\|\nabla v\|^p_{L^2(\Omega)} - c\,\|f\|_{L^2(\Omega)}\|\nabla v\|_{L^2(\Omega)}.$$

The last quantity converges to $+\infty$ if $\|\nabla v\|_{L^2(\Omega)} \to +\infty$, since $p < 2$. This means that J is coercive. Moreover the last inequality implies that J is bounded from below. Finally it is easy to verify that J satisfies the hypotheses of Theorem 9.2. By Theorem 9.1 J has a minimizer.

9.3 Euler equation

We will now give the definition of the Euler equation associated with a functional. The Euler equation is related to the Gâteaux differentiability of the functional. We will use the same notations as in Definition 9.29.

Definition 9.5. Let X be a Banach space. Let $J : X \to \mathbb{R}$ be a functional attaining its minimum at u. Suppose that J is Gâteaux differentiable. The equation $< J'(u), \varphi >= 0$, $\varphi \in X$, is the Euler equation associated with J.

We are now going to state a result that allows us to write the Euler equation associated with integral functionals. We will assume a p-growth in the gradient, with $p < N$. We refer to [27] for the case $p \geq N$ and for the proofs.

Theorem 9.6. *Let $j : \Omega \times \mathbb{R} \times \mathbb{R}^N \to \mathbb{R}$. Let*

$$J(v) = \int_\Omega j(x, v, \nabla v)$$

be defined on $W_0^{1,p}(\Omega)$. Assume that $j = j(x,s,\xi)$, $\frac{\partial j}{\partial s}$, $\nabla_\xi j$ are Carathéodory functions. Let u_0 be a minimizer for J. Then the Euler equation

$$\int_\Omega \nabla_\xi j(x, u_0, \nabla u_0) \cdot \nabla \varphi + \int_\Omega \frac{\partial j}{\partial s}(x, u_0, \nabla u_0)\varphi = 0$$

is satisfied for every $\varphi \in W_0^{1,p}(\Omega)$ if there exist $\alpha_1 \in L^1(\Omega)$, $\alpha_2 \in L^{\frac{Np}{Np-N+p}}(\Omega)$, $\alpha_3 \in L^{p'}(\Omega)$ and $\beta \geq 0$ such that

(1) $|j(x,s,\xi)| \leq \alpha_1(x) + \beta(|u|^{p^*} + |\xi|^p)$;

(2) $|\frac{\partial j}{\partial s}| \leq \alpha_2(x) + \beta(|u|^{r_1} + |\xi|^{r_2})$, $|\nabla_\xi j| \leq \alpha_3(x) + \beta(|u|^q + |\xi|^{p-1})$ *with* $r_1 \leq p^* - 1, r_2 \leq \frac{Np-N+p}{N}, q \leq \frac{Np-N}{N-p}$.

Remark 9.7. In the case where j satisfies condition 1 of the above theorem and the derivatives of j satisfy $\left|\frac{\partial j}{\partial s}\right| \leq \alpha_1(x) + \beta|\xi|^p$, $|\nabla_\xi j| \leq \alpha_2(x) + \beta|\xi|^{p-1}$ for some $\alpha_2 \in L^{p'}(\Omega)$, the Euler equation is verified for every $\varphi \in W_0^{1,p}(\Omega) \cap L^\infty(\Omega)$, provided the minimizer u_0 is bounded. We will see in the next section sufficient conditions assuring the boundedness of minimizers.

We are going to give some applications of the previous results.

Example 9.8. Let

$$J: \quad H_0^1(\Omega) \to \mathbb{R},$$

be defined by

$$J(v) = \frac{1}{2}\int_\Omega M(x)\nabla v \cdot \nabla v - \int_\Omega f v$$

where $M(x)$ is a $\mathbb{R}^{N\times N}$ matrix with bounded entries and such that $M(x)\xi\cdot\xi \geq \alpha|\xi|^2$ for some $\alpha > 0$; moreover $f \in L^2(\Omega)$. It is easy to see that J attains its minimum at u using Theorems 9.1 and 9.2. Note that the coercivity of J is easily obtained by Hölder's and Poincaré's inequalities:

$$J(v) \geq \frac{\alpha}{2}\|\nabla v\|^2_{L^2(\Omega)} - \|f\|_{L^2(\Omega)}\|v\|_{L^2(\Omega)} \geq \frac{\alpha}{2}\|\nabla v\|^2_{L^2(\Omega)} - c\,\|f\|_{L^2(\Omega)}\|\nabla v\|_{L^2(\Omega)}\,.$$

The Euler equation that u satisfies is

$$\begin{cases} -\operatorname{div}(M(x)\nabla u) = f\,, & \text{in } \Omega\,,\\ u = 0\,, & \text{on } \partial\Omega\,,\end{cases}$$

by Theorem 9.6.

With the same technique one can prove the existence of a minimizer $u \in H_0^1(\Omega)$ of

$$J(v) = \frac{1}{2}\int_\Omega M(x)\nabla v \cdot \nabla v - \int_\Omega f v + \int_\Omega G(v)\,,$$

where $G : \mathbb{R} \to \mathbb{R}$ is a C^1 function such that $0 \leq G(s) \leq \gamma_1|s|^{q_1}$, $q_1 \leq 2^*$ and $|g(s)| = |G'(s)| \leq \gamma_2|s|^{q_2}$, $q_2 \leq 2^* - 1$, with $\gamma_1, \gamma_2 \geq 0$. Therefore there exists a solution $u \in H_0^1(\Omega)$ to

$$\begin{cases} -\operatorname{div}(M(x)\nabla u) + g(u) = f\,, & \text{in } \Omega\,,\\ u = 0\,, & \text{on } \partial\Omega\,.\end{cases}$$

Example 9.9. Let

$$J: \quad H_0^1(\Omega) \to \mathbb{R},$$

be defined by

$$J(v) = \int_\Omega M(x)\nabla v \cdot \nabla v - \lambda_1 \int_\Omega v^2$$

where $M(x)$ is a $\mathbb{R}^{N\times N}$ matrix with bounded entries and such that $M(x)\xi\cdot\xi \geq \alpha|\xi|^2$ for some $\alpha > 0$; moreover λ_1 is the smallest eigenvalue of $L(v) = -\operatorname{div}(M(x)\nabla v)$. As a consequence of Theorem 8.4, J attains its minimum at $u \in H_0^1(\Omega)$, which satisfies

$$\begin{cases} -\operatorname{div}(M(x)\nabla u) = \lambda_1 u\,, & \text{in } \Omega\,,\\ u = 0\,, & \text{on } \partial\Omega\,,\end{cases}$$

by Theorem 9.6.

Remark 9.10. There exist however equations that are not the Euler equation associated with some functional, as we will see in Example 9.16.

In the next example we will solve a constrained minimization problem. We will see a different proof in Theorem 9.25.

Example 9.11. Let J be defined on $H_0^1(\Omega)$ by

$$J(v) = \frac{1}{2}\int_\Omega |\nabla v|^2 - \frac{1}{p}\int_\Omega |v|^p$$

with $2 < p < 2^*$. We claim that J does not have a minimizer in $H_0^1(\Omega)$. Indeed, if $\varphi_1 \in H_0^1(\Omega)$ is the eigenfunction corresponding to the first eigenvalue of the Laplacian operator (that is, $-\Delta\varphi_1 = \lambda_1\varphi_1$), one has, for $t > 0$,

$$J(t\varphi_1) = \frac{t^2}{2}\int_\Omega |\nabla\varphi_1|^2 - \frac{t^p}{p}\int_\Omega |\varphi_1|^p = \frac{t^2}{2}\lambda_1\int_\Omega \varphi_1^2 - \frac{t^p}{p}\int_\Omega |\varphi_1|^p .$$

The limit as $t \to +\infty$ proves that J is unbounded from below, since $p > 2$. Nevertheless we are going to prove that J has a minimizer on

$$A = \left\{ v \in H_0^1(\Omega) : \int_\Omega |v|^p = 1 \right\} .$$

It is clear that J is bounded from below on A, by Poincaré's inequality:

$$J(v) \geq \frac{c^2}{2}\|v\|_{L^2(\Omega)}^2 - \frac{1}{p}\|v\|_{L^p(\Omega)}^p \geq -\frac{1}{p} .$$

Moreover if v_n is a minimizing sequence, up to a subsequence, v_n converges weakly in $H_0^1(\Omega)$ to some function u in A, which is a minimizer, as J is weakly lower semicontinuous. Let us write the equation that u satisfies. We have that

$$J(u) \leq J\left(\frac{u + tv}{\|u + tv\|_{L^p(\Omega)}}\right) ,$$

for every $t \in \mathbb{R}$ and for every $v \in H_0^1(\Omega)$. Let us set

$$g(t) = J\left(\frac{u + tv}{\|u + tv\|_{L^p(\Omega)}}\right) .$$

We note that

$$\frac{d}{dt}\left(\int_\Omega |u + tv|^p\right) = p\int_\Omega |u + tv|^{p-2}(u + tv)v .$$

Using that $u \in A$ we obtain

$$g'(0) = \left[\int_\Omega \nabla u \cdot \nabla v - \int_\Omega |\nabla u|^2 \int_\Omega |u|^{p-2} u v \right] = 0 .$$

Therefore u satisfies

$$-\Delta u - \|u\|^2_{H_0^1(\Omega)} |u|^{p-2} u = 0 .$$

Now, let $w = tu$, for $t > 0$; then w satisfies

$$-\Delta w = \frac{t\|u\|^2_{H_0^1(\Omega)}}{t^{p-1}} |w|^{p-2} w .$$

Choosing t such that $\frac{t\|u\|^2_{H_0^1(\Omega)}}{t^{p-1}} = 1$ we infer the existence of a $H_0^1(\Omega)$ solution to $-\Delta w = |w|^{p-2} w$.

Example 9.12. We prove the existence of a $W_0^{1,p}(\Omega)$ solution to the following problem:

$$\begin{cases} -\operatorname{div}(|\nabla u|^{p-2} \nabla u) = b(x, u) , & \text{in} \quad \Omega , \\ u = 0 , & \text{on} \quad \partial\Omega , \end{cases} \tag{9.3.1}$$

where $p \geq 2$ and $b(x, s) : \Omega \times \mathbb{R} \to \mathbb{R}$ is a Carathéodory bounded function. We remark that this problem is a Leray–Lions problem (see Chapter 5). Here we want to show how it can be solved using the tools of the Calculus of Variations and Schauder's theorem.

Let us define $\sigma : W_0^{1,p}(\Omega) \to W_0^{1,p}(\Omega)$ as the map that associates with $w \in W_0^{1,p}(\Omega)$ the solution $z \in W_0^{1,p}(\Omega)$ of

$$\begin{cases} -\operatorname{div}(|\nabla z|^{p-2} \nabla z) = b(x, w) , & \text{in} \quad \Omega , \\ z = 0 , & \text{on} \quad \partial\Omega . \end{cases} \tag{9.3.2}$$

This map is well defined, since by Theorems 9.1 and 9.2

$$J(v) = \frac{1}{p} \int_\Omega |\nabla v|^p - \int_\Omega b(x, w) v$$

has a minimizer and the Euler equation is

$$-\operatorname{div}(|\nabla z|^{p-2} \nabla z) = b(x, w) .$$

Moreover such a minimizer is unique, by the strict convexity of J.

Let us prove that there exists a bounded, convex, invariant set for σ and that σ is completely continuous. The existence of a solution to problem (9.3.1) will follow from Theorem 2.10. Let us consider z in the weak formulation of problem (9.3.2) as a test function: by Poincaré's inequality, we have

$$\|\nabla z\|^p_{L^p(\Omega)} = \int_\Omega |\nabla z|^{p-2} \nabla z \cdot \nabla z = \int_\Omega b(x, w) z \leq C \|\nabla z\|_{L^p(\Omega)}$$

since b is bounded. This implies that there exists R such that if $\|\nabla w\|_{L^p(\Omega)} \le R$ then $\|\nabla z\|_{L^p(\Omega)} \le R$, that is, there exists a convex, bounded, invariant set for σ.

To prove that σ is completely continuous, it is sufficient to prove that if $w_n \to w$ weakly in $W_0^{1,p}(\Omega)$ then $\sigma(w_n) \to \sigma(w)$ in $W_0^{1,p}(\Omega)$. To this end, by choosing $z_n - z$ as a test function, we have

$$\int_\Omega |\nabla z_n|^{p-2}\nabla z_n \cdot \nabla(z_n - z) = \int_\Omega b(x, w_n)(z_n - z)$$

$$\int_\Omega |\nabla z|^{p-2}\nabla z \cdot \nabla(z_n - z) = \int_\Omega b(x, w)(z_n - z).$$

Subtracting side by side

$$\int_\Omega \Big[|\nabla z_n|^{p-2}\nabla z_n - |\nabla z|^{p-2}\nabla z\Big] \cdot \nabla(z_n - z) = \int_\Omega [b(x, w_n) - b(x, w)](z_n - z).$$

Now, let us use the inequality $[|s|^{p-2}s - |t|^{p-2}t](s-t) \ge C|s-t|^p$ for $s, t \in \mathbb{R}$ on the left-hand side and Hölder's inequality on the right-hand side. We obtain

$$C\int_\Omega |\nabla(z_n - z)|^p \le \|b(x, w_n) - b(x, w)\|_{L^{p'}(\Omega)}\|z_n - z\|_{L^p(\Omega)}.$$

Poincaré's inequality implies

$$C\|\nabla(z_n - z)\|_{L^p(\Omega)}^{p-1} \le \|b(x, w_n) - b(x, w)\|_{L^{p'}(\Omega)}.$$

By Theorem 3.6 $b(x, w_n) \to b(x, w)$ in $L^{p'}(\Omega)$, since $w_n \to w$ in $L^p(\Omega)$. Consequently $z_n \to z$ in $W_0^{1,p}(\Omega)$, that is, $\sigma(w_n) \to \sigma(w)$ in $W_0^{1,p}(\Omega)$.

9.4 Summability of minimizers of integral functionals

In this section we study the regularity of the minimizers of the functionals

$$J(v) = \int_\Omega j(x, v, \nabla v) - \int_\Omega fv, \quad v \in H_0^1(\Omega),$$

where $j : \Omega \times \mathbb{R} \times \mathbb{R}^N \to \mathbb{R}$ is a Carathéodory function, convex in the last variable, such that

$$j(x, s, \xi) \ge \alpha|\xi|^2 \tag{9.4.1}$$

for some positive α. Moreover we assume that $j(x, s, 0) = 0$. Observe that if f belongs to $L^{\frac{2N}{N+2}}(\Omega)$, the assumptions on j assure the existence of a minimizer $u \in H_0^1(\Omega)$ of J, by Theorems 9.1 and 9.2. We are going to present some regularity results on u, according to the summability of f, proving the following results (see [18]):

$$f \in L^m(\Omega), \qquad m > N/2 \Rightarrow u \in H_0^1(\Omega) \cap L^\infty(\Omega)$$

$$f \in L^m(\Omega), \qquad m \in (2N/(N+2), N/2) \Rightarrow u \in H_0^1(\Omega) \cap L^{m^{**}}(\Omega)$$

Remark 9.13. We observe that we have the same summability of the solutions to Leray–Lions problem (see Theorems 6.6 and 6.9).

Remark 9.14. In the case where f has a lower summability, we will introduce the notion of T-minimum in Chapter 11.

In the sequel we will use the following sets:

$$A_k = \{|u| \geq k\}, \quad B_k = \{k \leq |u| < k+1\}.$$

Theorem 9.15. *Let $u \in H_0^1(\Omega)$ be a minimizer of J. Let $f \in L^m(\Omega)$, with $m > \frac{N}{2}$. Then u is bounded.*

Proof. By the minimality of u, we have $J(u) \leq J(T_k(u))$, that is,

$$\int_\Omega j(x,u,\nabla u) \leq \int_\Omega j(x, T_k(u), \nabla T_k(u)) + \int_\Omega f\, G_k(u).$$

This is equivalent to

$$\int_{A_k} j(x,u,\nabla u) \leq \int_\Omega f G_k(u).$$

Assumption (9.4.1) on j implies that

$$\alpha \int_{A_k} |\nabla G_k(u)|^2 \leq \int_\Omega f G_k(u). \tag{9.4.2}$$

As observed in Remark 6.8 we deduce from estimate (9.4.2) that u is bounded. □

Example 9.16. Let $J : H_0^1(\Omega) \to \mathbb{R}$ be defined by

$$J(v) = \frac{1}{2}\int_\Omega a(v)|\nabla v|^2 - \int_\Omega fv \tag{9.4.3}$$

under the following hypotheses:
(1) $0 < \alpha \leq a(s) \leq \beta$, for some $\alpha, \beta > 0$;
(2) a is differentiable and there exists $\gamma > 0$ such that $|a'(s)| \leq \gamma$;
(3) f belongs to $L^{\frac{2N}{N+2}}(\Omega)$.

By the above result, J attains its minimum at $u \in H_0^1(\Omega) \cap L^\infty(\Omega)$ satisfying

$$\int_\Omega a(u)\nabla u \cdot \nabla v + \frac{1}{2}\int_\Omega a'(u)|\nabla u|^2 v - \int_\Omega fv = 0, \quad \forall\, v \in H_0^1(\Omega) \cap L^\infty(\Omega)$$

by Remark 9.7. Consequently the equation

$$-\mathrm{div}(a(u)\nabla u) + g(u)|\nabla u|^2 = f$$

is the Euler equation associated with a functional if and only if $g = a'/2$.

Theorem 9.17. *Let $u \in H_0^1(\Omega)$ be a minimizer of J. Let $f \in L^m(\Omega)$, with $\frac{2N}{N+2} < m < \frac{N}{2}$. Then $u \in L^{m^{**}}(\Omega)$.*

Proof. We are going to show the regularity of u in a recursive way.

Step I: As in the proof of Theorem 9.15, we have, by the minimality of u

$$\alpha \int_{A_k} |\nabla G_k(u)|^2 \le \int_{A_k} |f||G_k(u)|$$

(see inequality (9.4.2)). This implies that

$$\alpha \int_{A_k} |\nabla u|^2 \le \int_{A_k} |f||u|. \tag{9.4.4}$$

Let $\lambda > 1$ and let M be a positive integer. Multiplying inequality (9.4.4) by $(1+k)^{2\lambda-3}$ and summing up for k from 0 to M we have

$$\alpha \sum_{k=0}^{M} (1+k)^{2\lambda-3} \int_{A_k} |\nabla u|^2 \le \sum_{k=0}^{M} (1+k)^{2\lambda-3} \int_{A_k} |f|\,|u|. \tag{9.4.5}$$

We are going to study the right-hand side of (9.4.5). Since $A_k = \bigcup_{j=k}^{+\infty} B_j$, we obtain

$$\begin{aligned}\sum_{k=0}^{M} (1+k)^{2\lambda-3} \int_{A_k} |f||u| &= \sum_{k=0}^{M} (1+k)^{2\lambda-3} \sum_{h=k}^{\infty} \int_{B_h} |f||u| \\ &= \sum_{h=0}^{+\infty} \int_{B_h} |f||u| \sum_{k=0}^{T_M(h)} (1+k)^{2\lambda-3},\end{aligned}$$

by exchanging the summation order. We now observe that for any $h \in \mathbb{N}$ it results

$$\sum_{k=0}^{T_M(h)} (1+k)^{2\lambda-3} \le \frac{1}{2(\lambda-1)} (2+T_M(h))^{(\lambda-1)2}.$$

Moreover on B_h one has $(1+T_M(h))^{(\lambda-1)2} \le (1+|T_M(u)|)^{(\lambda-1)2}$. Therefore, the right-hand side of (9.4.5) can be estimated by

$$\begin{aligned}\sum_{k=0}^{M} (1+k)^{2\lambda-3} \int_{A_k} |f|\,|u| &\le \frac{1}{2(\lambda-1)} \sum_{h=0}^{\infty} \int_{B_h} |f||u|[2+|T_M(u)]^{(\lambda-1)2} \\ &\le \frac{1}{2(\lambda-1)} \int_{\Omega} |f||u|[2+|T_M(u)]^{(\lambda-1)2}.\end{aligned}$$

We handle the left-hand side of (9.4.5) in a similar way:

$$\begin{aligned}\sum_{k=0}^{M} (1+k)^{2\lambda-3} \int_{A_k} |\nabla u|^2 &= \sum_{k=0}^{M} (1+k)^{2\lambda-3} \sum_{h=k}^{\infty} \int_{B_h} |\nabla u|^2 \\ &= \sum_{h=0}^{+\infty} \int_{B_h} |\nabla u|^2 \sum_{k=0}^{T_M(h)} (1+k)^{2\lambda-3}.\end{aligned} \tag{9.4.6}$$

We observe that

$$\frac{[1+T_M(h)]^{(\lambda-1)2}}{2(\lambda-1)} \le \sum_{k=0}^{T_M(h)} (1+k)^{2\lambda-3}\,.$$

Moreover on B_h one has $|T_M(u)|^{(\lambda-1)2} \le (1+T_M(h))^{(\lambda-1)2}$. Therefore, by (9.4.6), the left-hand side can be estimated from below by

$$\begin{aligned}\frac{\alpha}{2(\lambda-1)}\int_\Omega |\nabla u|^2 |T_M(u)|^{(\lambda-1)2} &\le \alpha \sum_{h=0}^{+\infty}\int_{B_h}|\nabla u|^2 \sum_{k=0}^{T_M(h)}(1+k)^{2\lambda-3}\\ &= \alpha\sum_{k=0}^{M}(1+k)^{2\lambda-3}\int_{A_k}|\nabla u|^2\,.\end{aligned}$$

In conclusion, if λ is any real greater than 1, it holds

$$\alpha\int_\Omega |\nabla u|^2 |T_M(u)|^{(\lambda-1)2} \le \int_\Omega |f||u|(2+|T_M(u)|)^{(\lambda-1)2}\,. \tag{9.4.7}$$

Step II: We are going to prove that

$$|f|\,|u|^{(\lambda-1)2+1} \in L^1(\Omega)\,. \tag{9.4.8}$$

Let $\eta_0 = 2^*$. Since f belongs to $L^m(\Omega)$ and u belongs to $L^{\eta_0}(\Omega)$, by Hölder's inequality, (9.4.8) holds true if $\lambda > 0$ is such that $\frac{1}{m} + \frac{2\lambda-1}{\eta} = 1$, that is,

$$\lambda = \lambda(\eta) = \frac{1}{2}\left(\frac{\eta}{m'} - 1\right) + 1\,.$$

Define

$$\lambda_0 = \lambda(\eta_0) = \frac{1}{2}\left[\frac{2^*}{m'} - 1\right] + 1\,.$$

Thus (9.4.8) holds for $\lambda = \lambda_0$. We have $\lambda_0 > 1$, since $m > (2^*)'$. Letting M tend to infinity in (9.4.7), we get

$$\alpha\int_\Omega |\nabla u|^2\,|u|^{(\lambda_0-1)2} \le \int_\Omega |f|\,(2+|u|)^{(\lambda_0-1)2+1}\,.$$

The Sobolev inequality on the left-hand side implies

$$\left[\int_\Omega |u|^{\lambda_0 2^*}\right]^{\frac{2}{2^*}} \le C\int_\Omega |f|\,(2+|u|)^{(\lambda_0-1)2+1}\,, \tag{9.4.9}$$

where C denotes a positive constant depending on S, λ_0, α. Define, for η in $\mathbb{R}$,

$$\gamma(\eta) = \lambda(\eta)\,2^* = \frac{1}{2}\left(\frac{\eta}{m'} - 1\right)2^* + 2^*\,. \tag{9.4.10}$$

By the definition of λ_0 $(\lambda_0 - 1)2 + 1 = \frac{\eta_0}{m'}$; by using Hölder's inequality on (9.4.9), we have

$$\left[\int_\Omega |u|^{\gamma(\eta_0)}\right]^{\frac{2}{2^*}} \le C\left\{\|f\|_{L^1(\Omega)} + \|f\|_{L^m(\Omega)}\left[\int_\Omega |u|^{\eta_0}\right]^{\frac{1}{m'}}\right\}.$$

Let $\theta = \frac{2^*}{2m'}$, and observe that $0 < \theta < 1$ since $m < \frac{N}{2}$. We then have

$$\int_\Omega |u|^{\gamma(\eta_0)} \le C\left\{\|f\|_{L^1(\Omega)}^{\frac{2^*}{2}} + \|f\|_{L^m(\Omega)}^{\frac{2^*}{2}}\left[\int_\Omega |u|^{\eta_0}\right]^{\theta}\right\}. \tag{9.4.11}$$

Set $\eta_1 = \gamma(\eta_0)$; we remark that $\eta_0 < \eta_1 < m^{**}$. Thus, we deduce from (9.4.11) that u belongs to $L^{\eta_1}(\Omega)$.

Step III: Since $\eta_1 > \eta_0$, we have $\lambda(\eta_1) > 1$; arguing as in Step *II* we can say that (9.4.8) holds true with $\lambda_1 = \lambda(\eta_1)$. Thus, it is possible to pass to the limit in (9.4.7) as M tends to infinity obtaining (9.4.9) with $\lambda = \lambda_1$, that is,

$$\left[\int_\Omega |u|^{\lambda_1 2^*}\right]^{\frac{2}{2^*}} \le C\int_\Omega |f|\,(2 + |u|)^{(\lambda_1 - 1)2+1}.$$

The same passages as above yield the following inequality, by definition of $\gamma(\eta)$ (see (9.4.10)):

$$\int_\Omega |u|^{\gamma(\eta_1)} \le C\left\{\|f\|_{L^1(\Omega)}^{\frac{2^*}{2}} + \|f\|_{L^m(\Omega)}^{\frac{2^*}{2}}\left[\int_\Omega |u|^{\eta_1}\right]^{\theta}\right\}.$$

By induction we get, setting $\eta_{k+1} = \gamma(\eta_k)$,

$$\int_\Omega |u|^{\eta_{k+1}} \le C + C\left\{\int_\Omega |u|^{\eta_k}\right\}^{\theta}. \tag{9.4.12}$$

The positive constant C above depends on η_k, $\|f\|_{L^1(\Omega)}$, α, S and $\text{meas}(\Omega)$. Since $\eta_k < m^{**}$, as $m > 1$, we can assume that C does not depend on k.

Step IV: We first observe that η_k is an increasing sequence. Using Hölder's inequality with exponent $\frac{\eta_{k+1}}{\eta_k}$, we have

$$\int_\Omega |u|^{\eta_k} \le \left[\int_\Omega |u|^{\eta_{k+1}}\right]^{\frac{\eta_k}{\eta_{k+1}}} \text{meas}(\Omega)^{1-\frac{\eta_k}{\eta_{k+1}}}. \tag{9.4.13}$$

By Hölder's inequality, (9.4.12) and (9.4.13) give:

$$\int_\Omega |u|^{\eta_{k+1}} \le C + C\,\text{meas}(\Omega)^{\theta\left[1-\frac{\eta_k}{\eta_{k+1}}\right]}\left[\int_\Omega |u|^{\eta_{k+1}}\right]^{\theta\frac{\eta_k}{\eta_{k+1}}}$$

$$\le C + C\left[1 + \int_\Omega |u|^{\eta_{k+1}}\right]^{\theta\frac{\eta_k}{\eta_{k+1}}}$$

$$\le C + C\left[1 + \int_\Omega |u|^{\eta_{k+1}}\right]^{\theta}$$

$$\le C + C\left[\int_\Omega |u|^{\eta_{k+1}}\right]^{\theta}.$$

The last inequality implies that

$$\int_\Omega |u|^{\eta_{k+1}} \le C, \tag{9.4.14}$$

as $\theta < 1$. Since $\{\eta_k\}$ is an increasing and bounded sequence, it converges to some $\rho > 0$. By (9.4.10), the limit $\rho > 0$, as k tends to infinity, is such that

$$\rho = \frac{1}{2}\Big(\frac{\rho}{m'} - 1\Big)2^* + 2^*,$$

that is, $\rho = m^{**}$. Letting k tend to infinity in (9.4.14) we obtain that u belongs to $L^{m^{**}}(\Omega)$. □

9.5 The Ekeland variational principle

In the previous section, we have seen the importance of the minimizing sequences of a functional. In this section, we explain the Ekeland variational principle, which is a useful tool in studying their behavior (see [28]). Indeed it allows us in some sense to have a "good" minimizing sequence, whose elements have some minimizing properties.

Theorem 9.18 (Ekeland principle). *Let (X, d) be a complete metric space. Let $J : X \to \mathbb{R} \cup \{+\infty\}$ be a lower semicontinuous functional which is bounded from below. Let $\overline{u} \in X$ be such that*

$$J(\overline{u}) \le \inf_X J + \frac{1}{n}. \tag{9.5.1}$$

Then there exists $v \in X$ such that

$$J(v) \leq J(\overline{u}) \tag{9.5.2}$$

$$d(v, \overline{u}) \leq 1 \tag{9.5.3}$$

$$J(v) < J(w) + \tfrac{1}{n} d(v, w), \quad \forall \quad w \in X, \quad w \neq v. \tag{9.5.4}$$

Proof. Let us define by induction a sequence $u_k \subset X$. We set $u_1 = \overline{u}$. Assume we have defined $u_1, u_2, \ldots u_k$. Let

$$S_k = \left\{ w \in X : J(w) \leq J(u_k) - \tfrac{1}{n} d(u_k, w) \right\}.$$

S_k is not empty, since $u_k \in S_k$. By definition of infimum, there exists $u_{k+1} \in S_k$ such that

$$J(u_{k+1}) \leq \frac{1}{2} \left\{ J(u_k) + \inf_{S_k} J \right\}. \tag{9.5.5}$$

Let us prove that u_k is a Cauchy sequence. Since $u_{k+1} \in S_k$ then

$$\tfrac{1}{n} d(u_k, u_{k+1}) \leq J(u_k) - J(u_{k+1}). \tag{9.5.6}$$

By the triangle inequality

$$\frac{1}{n} d(u_k, u_{k+m}) \leq \frac{1}{n} \sum_{j=1}^{m} d(u_{k+j}, u_{k+j-1}) \leq J(u_k) - J(u_{k+m}). \tag{9.5.7}$$

Now, from (9.5.6) it follows that $J(u_k)$ is decreasing; since J is bounded from below in X, one has that

$$\lim_{k \to \infty} J(u_k) = \alpha$$

for some $\alpha \in \mathbb{R}$. Inequality (9.5.7) proves that u_k is a Cauchy sequence. Consequently there exists $v \in X$ such that $v = \lim_{k \to \infty} u_k$. On the other hand, since J is lower semicontinuous,

$$J(v) \leq \liminf_{m \to \infty} J(u_{k+m}) = \alpha.$$

This inequality and the limit for $m \to +\infty$ in (9.5.7) imply that

$$\frac{1}{n} d(u_k, v) \leq J(u_k) - J(v). \tag{9.5.8}$$

For $k = 1$ we have

$$\frac{1}{n} d(\overline{u}, v) \leq J(\overline{u}) - J(v) \leq J(\overline{u}) - \inf_X J \leq \frac{1}{n}$$

by hypothesis (9.5.1). Hence $d(\overline{u}, v) \leq 1$ and $J(v) \leq J(\overline{u})$, that is, (9.5.2) and (9.5.3) hold.

To prove (9.5.4), assume by contradiction that there exists $w \in X$ such that

$$J(w) < J(v) - \frac{1}{n} d(w, v). \tag{9.5.9}$$

Using (9.5.8) one has

$$J(w) < J(u_k) - \frac{1}{n}d(u_k, v) - \frac{1}{n}d(w, v) < J(u_k) - \frac{1}{n}d(u_k, w),$$

that is, $w \in S_k$, for every k. Therefore

$$\inf_{S_k} J \le J(w).$$

By (9.5.5) and (9.5.9) we obtain

$$2J(u_{k+1}) - J(u_k) \le J(w) < J(v) - \frac{1}{n}d(w, v).$$

Passing to the limit for $k \to \infty$ we get

$$J(v) \le J(w) < J(v) - \frac{1}{n}d(v, w)$$

which is a contradiction. □

Remark 9.19. Let us introduce in X the distance $d_1 = \sqrt{\frac{1}{n}}d$. Then (X, d_1) is a complete metric space. From Theorem 9.18 it follows that if u_n is a minimizing sequence, there exists $v_n \in X$ such that
(1) $J(v_n) \le J(u_n)$
(2) $d(u_n, v_n) \le \sqrt{\frac{1}{n}}$
(3) $J(v_n) \le J(w) + \sqrt{\frac{1}{n}}d(v_n, w), \quad \forall w \in X.$

Hence v_n is a minimizing sequence and its elements have some minimizing properties.

We now apply Ekeland's principle to study some regular functionals (see the appendix for the definitions).

Theorem 9.20. *Let $(X, \|\cdot\|)$ be a Banach space and let $J : X \to \mathbb{R}$ be a semicontinuous functional that is bounded from below. Let J be Gâteaux differentiable in every direction $w \in X$. Then for every $n > 0$ there exists $u_n \in X$ such that*

$$J(u_n) \le \inf_X J + \frac{1}{n}$$
$$\|J'(u_n)\|_{X'} \le \frac{1}{n}.$$

Proof. The first inequality follows from the definition of infimum. Let us prove the second one. Theorem 9.18 implies that there exists $u_n \in X$ such that

$$J(u_n) \le J(v) + \frac{1}{n}\|v - u_n\|, \quad \text{for all } v \in X.$$

Let $w \in X$ and $t > 0$ be fixed. Choosing $v = u_n + tw$ in the previous inequality, we get

$$\frac{J(u_n) - J(u_n + tw)}{t} \leq \frac{1}{n}\|w\|.$$

Passing to the limit as $t \to 0$ in the previous inequality, one has $\langle J'(u_n), w\rangle \leq \frac{1}{n}\|w\|$ for every $w \in X$. Since this equality is valid for w and $-w$ we get

$$|\langle J'(u_n), w\rangle| \leq \frac{1}{n}\|w\|, \qquad \forall\, w \in X.$$

Therefore

$$\|J'(u_n)\|_{X'} = \sup_{w\in X, w\neq 0} \frac{\langle J'(u_n), w\rangle}{\|w\|} \leq \frac{1}{n}.$$

□

Definition 9.21. Let $(X, \|\cdot\|)$ be a Banach space and $J : X \to \mathbb{R}$ be a C^1 functional. We say that J satisfies the Palais–Smale condition if every sequence u_n in X such that the sequence $|J(u_n)|$ is bounded and $J'(u_n) \to 0$ in X' has a converging subsequence.

We are going to prove the following theorem (see [41]):

Theorem 9.22 (Minimization with the Palais–Smale condition). *Let $(X, \|\cdot\|)$ be a Banach space and $J : X \to \mathbb{R}$ a C^1 functional satisfying the Palais–Smale condition. Assume that J is bounded from below. Then J attains its minimum at $u_0 \in X$ and u_0 is a critical point for J, that is, $J'(u_0) = 0$.*

Proof. By Theorem 9.20 for every n there exists $u_n \in X$ such that

$$J(u_n) \leq \inf_X J + \frac{1}{n}, \qquad \|J'(u_n)\|_{X'} \leq \frac{1}{n}.$$

The Palais–Smale condition implies the existence of a subsequence u_{n_j} and $u_0 \in X$ such that $u_{n_j} \to u_0$. From the continuity of J and J', passing to the limit as $n \to \infty$ we get

$$J(u_0) = \inf_X J, \quad J'(u_0) = 0.$$

□

We are going to give an application of the previous theorem to the study of the critical points of a functional.

Theorem 9.23. *Let λ_1 be the first eigenvalue of $L(v) = -\Delta v$. Let $f \in L^p(\Omega)$ with $2 < p < 2^*$. Then*

$$J(v) = \frac{1}{2}\int_\Omega |\nabla v|^2 - \frac{\lambda_1}{2}\int_\Omega v^2 + \frac{1}{p}\int_\Omega |v|^p - \int_\Omega fv$$

defined on $H_0^1(\Omega)$, attains its minimum at u which satisfies

$$u \in H_0^1(\Omega) : -\Delta u + |u|^{p-2}u = \lambda_1 u + f.$$

Proof. We will prove the existence of a minimizer using Theorem 9.22. J is a C^1 functional. Moreover it is bounded from below, because

$$\frac{1}{2}\int_\Omega |\nabla v|^2 - \frac{\lambda_1}{2}\int_\Omega v^2 \geq 0$$

as we have seen in Theorem 8.4. On the other hand, using Hölder's inequality we have that

$$\frac{1}{p}\int_\Omega |v|^p - \int_\Omega fv \geq \frac{1}{p}\int_\Omega |v|^p - \|f\|_{L^{p'}(\Omega)}\|v\|_{L^p(\Omega)}$$

which is bounded from below. Let us prove that J satisfies the Palais–Smale condition. Let u_n be a sequence such that

$$|J(u_n)| \leq R \tag{9.5.10}$$

for some $R > 0$ and

$$J'(u_n) \to 0\,. \tag{9.5.11}$$

We want to prove that u_n converges in $H_0^1(\Omega)$, up to subsequence. Inequality (9.5.10) is equivalent to

$$-R \leq \frac{1}{2}\int_\Omega |\nabla u_n|^2 - \frac{\lambda_1}{2}\int_\Omega u_n^2 + \frac{1}{p}\int_\Omega |u_n|^p - \int_\Omega fu_n \leq R\,.$$

Since

$$\frac{1}{2}\int_\Omega |\nabla u_n|^2 - \frac{\lambda_1}{2}\int_\Omega u_n^2 \geq 0$$

one has, using Young's inequality,

$$\frac{1}{p}\int_\Omega |u_n|^p \leq R + \int_\Omega fu_n \leq R + \frac{1}{2p}\int_\Omega |u_n|^p + c(p)\int_\Omega |f|^{p'}\,.$$

Consequently the sequence u_n is bounded in $L^p(\Omega)$ and so, up to a subsequence, $u_n \to u$ weakly in $L^p(\Omega)$. On the other hand (9.5.11) gives us

$$-\Delta u_n + |u_n|^{p-2}u_n - \lambda_1 u_n - f = y_n\,, \tag{9.5.12}$$

with $y_n \in H^{-1}(\Omega)$ converging to 0 in $H^{-1}(\Omega)$. Considering u_n as a test function, one has

$$\int_\Omega |\nabla u_n|^2 \leq \int_\Omega |\nabla u_n|^2 + \int_\Omega |u_n|^p = \lambda_1\int_\Omega u_n^2 + \int_\Omega fu_n + \langle y_n, u_n\rangle\,.$$

The right-hand side is uniformly bounded and so $u_n \to u$ weakly in $H_0^1(\Omega)$ and $u_n \to u$ in $L^p(\Omega)$, since $p < 2^*$. We claim that $u_n \to u$ in $H_0^1(\Omega)$. Choosing $u_n - u$ as a test function, (9.5.12) implies

$$\int_\Omega \nabla(u_n-u)\cdot\nabla(u_n-u) = -\int_\Omega |u_n|^{p-2}u_n(u_n-u) + \lambda_1\int_\Omega u_n(u_n-u) + \int_\Omega f(u_n-u) + \langle y_n, u_n-u\rangle - \int_\Omega \nabla u\cdot\nabla(u_n-u)\,.$$

It is easily seen that the right-hand side goes to zero, by the fact that $u_n \to u$ weakly in $H_0^1(\Omega)$ and strongly in $L^p(\Omega)$; therefore $u_n \to u$ in $H_0^1(\Omega)$. Due to the previous theorem, J attains its minimum at $u \in H_0^1(\Omega)$. The Euler equation is

$$u \in H_0^1(\Omega): \quad -\Delta u + |u|^{p-2}u = \lambda_1 u + f\,. \qquad \square$$

For functionals that are unbounded from below, the mountain pass theorem by Ambrosetti and Rabinowitz [1]) can be useful:

Theorem 9.24 (Mountain Pass theorem). *Let H be a Hilbert space. Let J be a C^1 functional defined on H satisfying the Palais–Smale condition and $J(0) = 0$. Assume that there exist positive constants r and a such that $J(u) \ge a$ if $\|u\| = r$, and there exists $v \in H$ with $\|v\| > r$ such that $J(v) \le 0$. Let*

$$\Gamma = \{\mathbf{g} \in C([0,1];H) \mid \mathbf{g}(0) = 0, \mathbf{g}(1) = v\}$$

Then

$$c = \inf_{\mathbf{g}\in\Gamma}\max_{0\le t\le 1} I[\mathbf{g}(t)]\,,$$

is a critical value of J.

Theorem 9.25. *The functional*

$$J(v) = \frac{1}{2}\int_\Omega |\nabla v|^2 - \frac{1}{p}\int_\Omega |v|^p\,, \quad 2 < p < 2^*\,,$$

defined on $H_0^1(\Omega)$, has a critical point at u which satisfies

$$u \in H_0^1(\Omega): \quad -\Delta u = |u|^{p-2}u\,.$$

Remark 9.26. We have already seen in Example 9.11 that J is unbounded from below.

Proof. We are going to prove that J satisfies the hypotheses of Theorem 9.24. This will imply the existence of a function $u \neq 0$ in $H_0^1(\Omega)$ such that $-\Delta u = |u|^{p-2}u$. It is not difficult to prove that J satisfies the Palais–Smale condition. Indeed let u_n be a sequence such that

$$|J(u_n)| \le R \tag{9.5.13}$$

for some $R > 0$ and

$$J'(u_n) \to 0\,. \tag{9.5.14}$$

We want to prove that u_n converges in $H_0^1(\Omega)$, up to subsequence. Choosing u_n as a test function in (9.5.14) we get

$$\int_\Omega |u_n|^p \le \|u_n\|_{H_0^1(\Omega)} + \|u_n\|^2_{H_0^1(\Omega)}.$$

On the other hand, (9.5.13) gives

$$\frac{1}{2}\|u_n\|_{H_0^1(\Omega)} \le R + \frac{1}{p}\int_\Omega |u_n|^p.$$

The last two estimates give that $\|u_n\|_{H_0^1(\Omega)}$ is uniformly bounded, since $p > 2$. Therefore there exists $u \in H_0^1(\Omega)$ such that, up to a subsequence, $u_n \to u$ weakly in $H_0^1(\Omega)$ and $u_n \to u$ in $L^p(\Omega)$, as $p < 2^*$. We claim that $u_n \to u$ in $H_0^1(\Omega)$. Indeed, (9.5.14) gives, for $y_n \in H^{-1}(\Omega)$ converging to 0 in $H^{-1}(\Omega)$

$$\begin{aligned}\int_\Omega \nabla(u_n - u)\cdot\nabla(u_n - u) = &\int_\Omega |u_n|^{p-2}u_n(u_n - u)\\ &+ \langle y_n, u_n - u\rangle - \int_\Omega \nabla u\cdot\nabla(u_n - u).\end{aligned}$$

It is easily seen that the right-hand side goes to zero, by the fact that $u_n \to u$ weakly in $H_0^1(\Omega)$ and strongly in $L^p(\Omega)$; hence $u_n \to u$ in $H_0^1(\Omega)$. Therefore J satisfies the Palais–Smale condition.

Now, let $u \in H_0^1(\Omega)$, with $\|u\|_{H_0^1(\Omega)} = r$. Then $J(u) = \frac{r^2}{2} - \frac{1}{p}\int_\Omega |u|^p$. By the Hölder and the Sobolev inequalities one has

$$\int_\Omega |u|^p \le \operatorname{meas}(\Omega)^{1-\frac{p}{2^*}}\left[\int_\Omega |u|^{2^*}\right]^{\frac{p}{2^*}} \le \frac{\operatorname{meas}(\Omega)^{1-\frac{p}{2^*}}}{S^p}\|u\|^p_{H_0^1(\Omega)} \le \frac{\operatorname{meas}(\Omega)^{1-\frac{p}{2^*}}}{S^p} r^p$$

and then $J(u) \ge \frac{r^2}{2} - \frac{\operatorname{meas}(\Omega)^{1-\frac{p}{2^*}}}{S^p} r^p = a > 0$ if r is sufficiently small, as $p > 2$. Now, let $u \in H_0^1(\Omega)$ be fixed and let $v = tu$, $t > 0$. Then

$$J(v) = \frac{t^2}{2}\int_\Omega |\nabla u|^2 - \frac{t^p}{p}\int_\Omega |u|^p < 0$$

if t is sufficiently large, as $p > 2$. □

Remark 9.27. For the regularity of minimizing sequences for integral functionals of the Calculus of Variations via the Ekeland's principle see [11].

9.6 Appendix

Let X be a Banach space and $\|x\|$ denote the norm of $x \in X$.

Definition 9.28. Let $J : X \to \mathbb{R}$ be a functional.
(1) J is weakly lower semicontinuous if $\liminf_{n\to\infty} J(x_n) \geq J(x)$ for every sequence x_n weakly converging to x.
(2) J is coercive if $\lim_{\|x\|\to+\infty} J(x) = +\infty$.

Let us recall the Gâteaux differentiability of a functional $J : X \to \mathbb{R}$ (see [2] for more details).

Definition 9.29. J is Gâteaux differentiable at $x \in X$ in the direction $h \in X$ if there exists a linear continuous functional $J'(x) : X \to \mathbb{R}$ such that

$$\lim_{t\to 0} \frac{J(x+th) - J(x)}{t} = \langle J'(x), h\rangle .$$

Let us recall the Fréchet differentiability.

Definition 9.30. J is Fréchet differentiable at $x \in X$ if there exists $A_x \in X'$ such that

$$\lim_{\|h\|\to 0} \frac{\|J(x+h) - J(x) - A_x(h)\|}{\|h\|} = 0 .$$

If the map

$$\begin{aligned} X &\to X' \\ x &\to A_x \end{aligned}$$

is continuous, we will say that J is C^1.

We note that if J is Fréchet differentiable, then it is Gâteaux differentiable. Conversely, if the map $x \to J'(x)$ is defined in a neighborhood of x_0 and is continuous at x_0, then J is Fréchet differentiable at x_0 and $J'(x_0) = A_{x_0}$.

Part II

10 Natural growth problems

10.1 Introduction

In Chapter 9, we have seen that a minimizer of

$$F(v) = \int_{\Omega} a(v)|\nabla v|^2 - \int_{\Omega} fv$$

satisfies the Euler equation

$$-\text{div}(a(u)\nabla u) + a'(u)|\nabla u|^2 = f$$

in which there appears the natural growth term $|\nabla u|^2$. In this chapter we are going to study the boundary value problem (not necessarily variational)

$$\begin{cases} -\text{div}(M(x,u)\nabla u) + \mu u = b(x,u,\nabla u) + f(x), & \text{in } \Omega, \\ u = 0, & \text{on } \partial\Omega, \end{cases} \tag{10.1.1}$$

under the following assumptions. The set Ω is an open bounded subset of $\mathbb{R}^N$, with $N \geq 3$. Moreover $b : \Omega \times \mathbb{R} \times \mathbb{R}^N \to \mathbb{R}$ is a Carathéodory function such that, for some $\gamma > 0$

$$|b(x,s,\xi)| \leq \gamma|\xi|^2; \tag{10.1.2}$$

$M = M(x,s)$ is a symmetric matrix with Carathéodory entries satisfying

$$M(x,s)\xi \cdot \xi \geq \alpha|\xi|^2, \quad |M(x,s)| \leq \beta. \tag{10.1.3}$$

Finally $\mu > 0$. Observe that, due to the quadratic growth of b with respect to the gradient, Leray–Lions theorem cannot be applied, since the composition operator $v \to b(x,v(x),\nabla v(x))$ does not map $H_0^1(\Omega)$ to its dual, but only to $L^1(\Omega)$. We refer to [20] for more details and for further references.

As well, a natural growth term appears in the case where one considers the minimization of integral functionals as

$$I(v) = \frac{1}{2}\int_{\Omega}(1+|v|^r)|\nabla v|^2 - \int_{\Omega} f\,v\,, v \in H_0^1(\Omega)\,.$$

Indeed, the Euler equation for I is (at least formally)

$$\begin{cases} -\text{div}((1+|u|^r)\nabla u) + \frac{r}{2}u|u|^{r-2}|\nabla u|^2 = f, & \text{in } \Omega, \\ u = 0, & \text{on } \partial\Omega. \end{cases} \tag{10.1.4}$$

Note that the lower order term depends quadratically on the gradient and satisfies $v\,g(x,v,\nabla v) \geq 0$. In the second part of this chapter we will study the boundary value problem

$$\begin{cases} -\text{div}([a(x)+|u|^q]\nabla u) + b(x)u|u|^{p-1}|\nabla u|^2 = f, & \text{in } \Omega, \\ u = 0, & \text{on } \partial\Omega, \end{cases} \tag{10.1.5}$$

under the following assumptions. The functions a and b are measurable and

$$0 < \alpha \le a(x) \le \beta, \tag{10.1.6}$$

$$0 < \mu \le b(x) \le \nu, \tag{10.1.7}$$

where α, β, μ, ν are fixed real numbers, and p, q are positive.

The case $p = q - 1$ is related with the minimization problems. We refer to [6] for more details and further references.

10.2 A problem with bounded solutions

We will present the following existence and regularity result.

Theorem 10.1. *Let Ω be an open bounded subset of $\mathbb{R}^N$, with $N \ge 3$. Let $\mu > 0$. Assume (10.1.2) and (10.1.3). Let f be in $L^m(\Omega)$, $m > \frac{N}{2}$. Then there exists a solution $u \in H_0^1(\Omega) \cap L^\infty(\Omega)$ to problem (10.1.1) in the following weak sense:*

$$\int_\Omega M(x,u)\nabla u\cdot\nabla\varphi + \mu\int_\Omega u\varphi = \int_\Omega b(x,u,\nabla u)\varphi + \int_\Omega f\varphi, \quad \forall\, \varphi \in H_0^1(\Omega)\cap L^\infty(\Omega).$$

We will work by approximation on the following problems:

$$\begin{cases} -\operatorname{div}(M(x,u)\nabla u) + \mu u = b_n(x,u,\nabla u) + f_n(x), & \text{in } \Omega, \\ u = 0, & \text{on } \partial\Omega, \end{cases} \tag{10.2.1}$$

where

$$b_n(x,s,\xi) = \frac{b(x,s,\xi)}{1 + \frac{1}{n}|b(x,s,\xi)|}$$

and

$$f_n(x) = \frac{f(x)}{1 + \frac{1}{n}|f(x)|}.$$

Corollary 5.10 of the Leray–Lions theorem and Stampacchia's theorem (Theorem 6.7) imply that for every $n \in \mathbb{N}$ there exists a weak solution u_n in $H_0^1(\Omega) \cap L^\infty(\Omega)$, since $|b_n| \le n$. The first step of the proof consists in getting uniform estimates of the $H_0^1(\Omega)$ and $L^\infty(\Omega)$ norms of the solutions; we therefore may pass to the limit and prove the existence of a solution to problem (10.1.1). The following lemmata will be useful for us.

Lemma 10.2. *Let u_n be the sequence of solutions to problems (10.2.1). Then the sequence u_n is bounded in $H_0^1(\Omega)$.*

Proof. Let us consider $\varphi_n = (e^{2\lambda|u_n|} - 1)\mathrm{sgn}(u_n)$, $\lambda > \frac{\gamma}{2\alpha}$, as a test function in the weak formulation of problems (10.2.1): this is possible since $u_n \in L^\infty(\Omega)$.

Step I: As for the left-hand side, using the ellipticity of M, one has

$$\int_\Omega M(x,u_n)\nabla u_n\cdot\nabla\varphi_n+\mu\int_\Omega u_n\varphi_n \geq 2\alpha\lambda\int_\Omega |\nabla u_n|^2 e^{2\lambda|u_n|}+\mu\int_\Omega |u_n|(e^{2\lambda|u_n|}-1). \tag{10.2.2}$$

Regarding the right-hand side, the growth hypothesis on b allows us to say that

$$\int_\Omega b(x,u_n,\nabla u_n)\varphi_n + \int_\Omega f_n\varphi_n \leq \int_\Omega \gamma\,|\nabla u_n|^2 e^{2\lambda|u_n|} + \int_\Omega |f|(e^{2\lambda|u_n|}-1).$$

It is easy to prove that

$$\int_\Omega |\nabla u_n|^2 e^{2\lambda|u_n|} = \frac{1}{\lambda^2}\int_\Omega |\nabla(e^{\lambda|u_n|}-1)|^2.$$

Therefore, by (10.2.2), we obtain

$$\frac{2\alpha\lambda-\gamma}{\lambda^2}\|\nabla(e^{\lambda|u_n|}-1)\|^2_{L^2(\Omega)} + \mu\int_\Omega |u_n|(e^{2\lambda|u_n|}-1) \leq \int_\Omega |f|\,(e^{2\lambda|u_n|}-1).$$

Step II: Let us now estimate the last term of the above inequality. Let $R > 1$. Using that $e^{2t}-1 \leq R\,(e^t-1)^2 + \frac{1}{R-1}$, $t \in \mathbb{R}^+$ and Hölder's inequality we have

$$\int_\Omega |f|\,(e^{2\lambda|u_n|}-1) \leq \frac{1}{R-1}\int_\Omega |f| + R\|f\|_{L^m(\Omega)}\|e^{\lambda|u_n|}-1\|^2_{L^{2m'}(\Omega)}.$$

We note that $2 < 2m' < \frac{2N}{N-2} = 2^*$. The interpolation inequality gives

$$\int_\Omega |f|\,(e^{2\lambda|u_n|}-1) \leq \frac{1}{R-1}\int_\Omega |f| + R\|f\|_{L^m(\Omega)}\|e^{\lambda|u_n|}-1\|^{2\theta}_{L^{2^*}(\Omega)}\|e^{\lambda|u_n|}-1\|^{2(1-\theta)}_{L^2(\Omega)},$$

where θ satisfies $\frac{1}{2m'} = \frac{\theta}{2^*} + \frac{1-\theta}{2}$. Young's inequality on the right-hand side implies

$$\begin{aligned}&\int_\Omega |f|\,(e^{2\lambda|u_n|}-1)\\ &\quad\leq \frac{1}{R-1}\int_\Omega |f| + \varepsilon(R\|f\|_{L^m(\Omega)})^{\frac{1}{\theta}}\|e^{\lambda|u_n|}-1\|^2_{L^{2^*}(\Omega)} + C_\varepsilon\|e^{\lambda|u_n|}-1\|^2_{L^2(\Omega)},\end{aligned}$$

where $C_\varepsilon = \varepsilon^{\frac{\theta}{1-\theta}}$. Summarizing we have obtained

$$\begin{aligned}&\frac{2\alpha\lambda-\gamma}{\lambda^2}\|\nabla(e^{\lambda|u_n|}-1)\|^2_{L^2(\Omega)} + \mu\int_\Omega |u_n|(e^{2\lambda|u_n|}-1)\\ &\quad\leq \frac{1}{R-1}\int_\Omega |f| + \varepsilon(R\|f\|_{L^m(\Omega)})^{\frac{1}{\theta}}\|e^{\lambda|u_n|}-1\|^2_{L^{2^*}(\Omega)} + C_\varepsilon\|e^{\lambda|u_n|}-1\|^2_{L^2(\Omega)}.\end{aligned}$$

Step III: Let us use Sobolev's inequality on the right-hand side and afterward choose ε in such a way that

$$\frac{2\lambda\alpha - \gamma}{2\lambda^2} = \frac{1}{S^2}\varepsilon(R\|f\|_{L^m(\Omega)})^{\frac{1}{\theta}}.$$

This implies that

$$\frac{2\alpha\lambda - \gamma}{2\lambda^2}\|\nabla(e^{\lambda|u_n|} - 1)\|^2_{L^2(\Omega)} + \mu\int_\Omega |u_n|(e^{2\lambda|u_n|} - 1) \leq \frac{1}{R-1}\int_\Omega |f| + C_\varepsilon\int_\Omega (e^{\lambda|u_n|} - 1)^2. \quad (10.2.3)$$

Since $(e^{\lambda t} - 1)^2 \leq e^{2\lambda t} - 1$, for $t \geq 0$, we have

$$\frac{2\alpha\lambda - \gamma}{2\lambda^2}\|\nabla(e^{\lambda|u_n|} - 1)\|^2_{L^2(\Omega)} + \mu\int_{\{C_\varepsilon \leq \mu|u_n|\}} |u_n|(e^{2\lambda|u_n|} - 1) \leq \frac{1}{R-1}\int_\Omega |f| + C_\varepsilon\int_{\{C_\varepsilon \leq \mu|u_n|\}} (e^{2\lambda|u_n|} - 1) + C_\varepsilon\int_{\{C_\varepsilon > \mu|u_n|\}} (e^{2\lambda|u_n|} - 1).$$

Consequently

$$\frac{2\alpha\lambda - \gamma}{2\lambda^2}\|\nabla(e^{\lambda|u_n|} - 1)\|^2_{L^2(\Omega)} \leq \frac{\|f\|_{L^1(\Omega)}}{R-1} + C_\varepsilon(e^{2\lambda\frac{C_\varepsilon}{\mu}} - 1)\,\mathrm{meas}(\Omega).$$

The last inequality tells us that the sequence $\int_\Omega e^{2\lambda|u_n|}|\nabla u_n|^2$ is bounded. As $\lambda > 0$, the sequence $\int_\Omega |\nabla u_n|^2$ is bounded. □

Lemma 10.3. *Let u_n be the sequence of solutions to problems* (10.2.1). *Then the sequence u_n is bounded in $L^\infty(\Omega)$.*

Proof. Let us consider $v_n = (e^{2\lambda|G_k(u_n)|} - 1)\mathrm{sgn}(u_n)$ as a test functions in problems (10.2.1). Since $G_k(u_n) = 0$ in $\{|u_n| \leq k\}$, we have

$$\int_\Omega M(x, u_n)\nabla u_n \cdot \nabla v_n + \mu\int_{\{|u_n| \geq k\}} u_n v_n = \int_{\{|u_n| \geq k\}} b(x, u_n, \nabla u_n)v_n + \int_{\{|u_n| \geq k\}} f_n v_n.$$

Using exactly the same arguments used in the previous lemma to get (10.2.3) we have

$$\frac{2\lambda\alpha - \gamma}{2\lambda^2}\|\nabla(e^{\lambda|G_k(u_n)|} - 1)\|^2_{L^2(\Omega)} + \mu\int_{\{|u_n| \geq k\}} |u_n|(e^{2\lambda|G_k(u_n)|} - 1) \leq \frac{1}{R-1}\int_{\{|u_n| \geq k\}} |f| + C_\varepsilon\int_{\{|u_n| \geq k\}} (e^{\lambda|G_k(u_n)|} - 1)^2.$$

By Sobolev's inequality on the left-hand side we deduce that

$$\frac{(2\lambda\alpha-\gamma)S^2}{2\lambda^2}\|e^{\lambda|G_k(u_n)|}-1\|^2_{L^{2^*}(\Omega)}+\mu\int\limits_{\{|u_n|\geq k\}}|u_n|(e^{2\lambda|G_k(u_n)|}-1)$$
$$\leq\frac{1}{R-1}\int\limits_{\{|u_n|\geq k\}}|f|+C_\varepsilon\int\limits_{\{|u_n|\geq k\}}(e^{\lambda|G_k(u_n)|}-1)^2.$$

The inequality $(e^{\lambda t}-1)^2\leq e^{2\lambda t}-1$, for $t\geq 0$ on the second term of the left-hand side gives

$$\frac{1}{2}\frac{(2\lambda\alpha-\gamma)S^2}{\lambda^2}\|e^{\lambda|G_k(u_n)|}-1\|^2_{L^{2^*}(\Omega)}+\mu\int\limits_{\{|u_n|\geq k\}}k(e^{\lambda|G_k(u_n)|}-1)^2$$
$$\leq\frac{1}{R-1}\int\limits_{\{|u_n|\geq k\}}|f|+C_\varepsilon\int\limits_{\{|u_n|\geq k\}}(e^{\lambda|G_k(u_n)|}-1)^2.$$

Choosing $k\geq\frac{C_\varepsilon}{\mu}$, we get

$$\frac{1}{2}\frac{(2\lambda\alpha-\gamma)S^2}{\lambda^2}\|e^{\lambda|G_k(u_n)|}-1\|^2_{L^{2^*}(\Omega)}\leq\frac{1}{R-1}\int\limits_{\{|u_n|\geq k\}}|f|.$$

On the left-hand side we use that $e^t-1\geq t$ for every $t\geq 0$ and Hölder's inequality with exponent 2^*. This gives

$$\lambda\int\limits_\Omega|G_k(u_n)|\leq\int\limits_\Omega|e^{\lambda|G_k(u_n)|}-1|\leq\text{meas}(\{|u_n|>k\})^{\frac{N+2}{2N}}\|e^{\lambda|G_k(u_n)|}-1\|_{L^{2^*}(\Omega)}$$
$$\leq C_1\,\text{meas}(\{|u_n|>k\})^{\frac{N+2}{2N}}\Bigg(\int\limits_{\{|u_n|\geq k\}}|f|\Bigg)^{\frac{1}{2}}$$

where C_1 denotes a constant depending on λ, α, S and R. From Hölder's inequality with exponent m on the right-hand side we get, for $k\geq\frac{C_\varepsilon}{\mu}$,

$$\lambda\int\limits_\Omega|G_k(u_n)|\leq C_1\,\|f\|^{\frac{1}{2}}_{L^m(\Omega)}\,\text{meas}(\{|u_n|>k\})^{\frac{N+2}{2N}+\frac{1}{2m'}}.$$

Remark 6.3 and Lemma 6.2 imply that $\|u_n\|_{L^\infty(\Omega)}$ is bounded. □

We can now prove the existence theorem of this section.

Proof. We divide the proof into two steps.

Step I: Let u_n be the solutions to problems (10.2.1). The previous lemmata imply the existence of a function $u \in H_0^1(\Omega)$ such that $\nabla u_n \to \nabla u$ weakly in $L^2(\Omega)$, up to a subsequence. We are going to prove that $u_n \to u$ in $H_0^1(\Omega)$. To this end, let us consider as a test function in problems (10.2.1) $v_n = \psi(u_n - u)$, where $\psi(t) = (e^{\lambda|t|} - 1)\mathrm{sgn}(t)$. We can write

$$\begin{aligned}
&\int_\Omega M(x, u_n)(\nabla u_n - \nabla u) \cdot (\nabla u_n - \nabla u)\psi'(u_n - u) \\
&\quad = -\mu \int_\Omega u_n v_n + \int_\Omega b_n(u_n, \nabla u_n) v_n \\
&\qquad - \int_\Omega M(x, u_n)\nabla u \cdot (\nabla u_n - \nabla u)\psi'(u_n - u) + \int_\Omega f_n v_n \\
&\quad \le - \int_\Omega M(x, u_n)\nabla u \cdot \nabla(u_n - u)\psi'(u_n - u) \\
&\qquad - \mu \int_\Omega u_n v_n + \gamma \int_\Omega |\nabla u_n|^2 |v_n| + \int_\Omega |f||v_n|.
\end{aligned}$$

Using the ellipticity of M, one gets

$$\begin{aligned}
\alpha \int_\Omega |\nabla u_n - \nabla u|^2 \psi'(u_n - u)
&\le - \int_\Omega M(x, u_n)\nabla u \cdot \nabla(u_n - u)\psi'(u_n - u) \\
&\quad - \mu \int_\Omega u_n v_n + \gamma \int_\Omega |\nabla u_n|^2 |v_n| + \int_\Omega |f||v_n|.
\end{aligned}$$

Now, $|\nabla u_n|^2 = |(\nabla u_n - \nabla u) + \nabla u|^2 \le 2|\nabla u_n - \nabla u|^2 + 2|\nabla u|^2$: this implies that

$$\begin{aligned}
&\int_\Omega |\nabla u_n - \nabla u|^2 [\alpha\psi'(u_n - u) - 2\gamma|\psi(u_n - u)|] \\
&\le 2\gamma \int_\Omega |\nabla u|^2 v_n - \int_\Omega M(x, u_n)\nabla u \cdot (\nabla u_n - \nabla u)\psi'(u_n - u) - \mu \int_\Omega u_n v_n + \int_\Omega |f||v_n|.
\end{aligned}$$

Observe that if $\lambda > \frac{2\gamma}{\alpha}$ then $\alpha\psi'(u_n - u) - 2\gamma|\psi(u_n - u)| \ge 1$. With this choice, setting $C = \sup_n \|\psi'(u_n - u)\|_{L^\infty(\Omega)}$, one has

$$\begin{aligned}
\|\nabla u_n - \nabla u\|^2_{L^2(\Omega)} &\le C \int_\Omega M(x, u_n)\nabla u \cdot (\nabla u_n - \nabla u) - \mu \int_\Omega u_n v_n \\
&\quad + 2\gamma \int_\Omega |\nabla u|^2 |v_n| + \int_\Omega |f||v_n|.
\end{aligned} \tag{10.2.4}$$

We now want to prove that the right-hand side of the previous inequality goes to 0. It is clear that $v_n \to 0$ a.e. in Ω and that $|v_n| \le C$ since ψ is continuous and u_n, u are bounded in $L^\infty(\Omega)$; Lebesgue's theorem implies then $\int_\Omega |\nabla u|^2 |v_n| \to 0$ and $\int_\Omega |f||v_n| \to 0$. Hölder's inequality with exponent 2 implies that $\int_\Omega u_n v_n \to 0$. Let us now study the first term. One has that $M(x, u_n)\nabla u \to M(x, u)\nabla u$ a.e. and consequently in $L^2(\Omega)$ for Lebesgue's theorem. Since $\nabla u_n \to \nabla u$ weakly in $L^2(\Omega)$, the first term goes to 0. Therefore (10.2.4) implies that $\|\nabla u_n - \nabla u\|_{L^2(\Omega)} \to 0$.

Step II: We now prove that u is a solution. For every $\varphi \in H_0^1(\Omega) \cap L^\infty(\Omega)$ one has

$$\int_\Omega M(x, u_n)\nabla u_n \cdot \nabla\varphi + \mu \int_\Omega u_n \varphi = \int_\Omega b_n(x, u_n, \nabla u_n)\varphi + \int_\Omega f_n \varphi .$$

The first term tends to

$$\int_\Omega M(x, u)\nabla u \cdot \nabla\varphi ,$$

as $u_n \to u$ in $H_0^1(\Omega)$ and $M(x, u_n)\nabla v \to M(x, u)\nabla v$ in $L^2(\Omega)$ for every $v \in H_0^1(\Omega)$. Clearly $\int_\Omega f_n\varphi \to \int_\Omega f\varphi$ and $\int_\Omega u_n\varphi \to \int_\Omega u\varphi$. We claim that $\int_\Omega b_n(x, u_n, \nabla u_n)\varphi \to \int_\Omega b(x, u, \nabla u)\varphi$. The sequences u_n and ∇u_n converge a.e. in Ω to u and ∇u respectively, and so $b_n(x, u_n, \nabla u_n) \to b(x, u, \nabla u)$ a.e.. Moreover $|b_n(x, u_n, \nabla u_n)| \le \gamma |\nabla u_n|^2$: this last sequence converges in $L^1(\Omega)$. Due to Lebesgue's theorem

$$b_n(x, u_n, \nabla u_n)\varphi \to b(x, u, \nabla u)\varphi$$

in $L^1(\Omega)$, up to a subsequence. We can therefore pass to the limit and get

$$\int_\Omega M(x, u)\nabla u \cdot \nabla\varphi + \mu \int_\Omega u\varphi = \int_\Omega b(x, u, \nabla u)\varphi + \int_\Omega f\varphi$$

for every $\varphi \in H_0^1(\Omega) \cap L^\infty(\Omega)$. □

Remark 10.4. The μu term, with $\mu > 0$, has an important role for the existence of solutions. Indeed, let us consider

$$\begin{cases} -\Delta u = |\nabla u|^2 + f, & \text{in } \Omega, \\ u = 0, & \text{on } \partial\Omega, \end{cases}$$

where f is bounded and strictly positive. By contradiction, let $u \in H_0^1(\Omega) \cap L^\infty(\Omega)$ be a solution to the previous problem. By the Maximum Principle (Theorem 4.12) u is positive. Let us set $z = e^u - 1$. Then z belongs to $H_0^1(\Omega) \cap L^\infty(\Omega)$ and is positive in Ω. Consequently

$$\nabla z = (z+1)\nabla u, \quad -\Delta z = -(z+1)[\Delta u + |\nabla u|^2] = f(z+1)$$

in the distributional sense. Choosing φ_1, the first eigenfunction of $L(v) = -\Delta v$ (see Chapter 8), as a test function, one gets

$$0 \leq \lambda_1 \int_\Omega \varphi_1 z = \int_\Omega \nabla\varphi_1 \cdot \nabla z = -\int_\Omega f(z+1)\varphi_1 < -\int_\Omega f z \varphi_1 \leq 0\,,$$

that is, a contradiction.

10.3 A problem with unbounded solutions

In this section we study problem (10.1.5), proving the existence of weak $H_0^1(\Omega)$ solutions. In the first result, we will study the case where the source f belongs to $L^m(\Omega)$, with $m > \frac{N}{2}$.

Theorem 10.5. *Let Ω be a bounded open subset of $\mathbb{R}^N$, $N \geq 3$. Assume* (10.1.6) *and* (10.1.7). *Let f be a positive function belonging to $L^m(\Omega)$, $m > \frac{N}{2}$. Then there exists a positive solution $u \in H_0^1(\Omega) \cap L^\infty(\Omega)$ to*

$$\begin{cases} -div([a(x) + u^q]\nabla u) + b(x)u^p|\nabla u|^2 = f\,, & in\ \ \Omega\,, \\ u = 0\,, & on\ \partial\Omega\,. \end{cases} \tag{10.3.1}$$

Proof. Step I: Let $u_n \in H_0^1(\Omega) \cap L^\infty(\Omega)$ be a solution of the Dirichlet problem

$$\begin{cases} -\operatorname{div}([a(x) + |T_n(u_n)|^q]\nabla u_n) & \\ \quad + b(x)T_n(u_n)|T_n(u_n)|^{p-1}|\nabla u_n|^2 + \dfrac{1}{n}u_n = f\,, & \text{in}\ \Omega\,, \\ u_n = 0\,, & \text{on}\ \partial\Omega\,. \end{cases} \tag{10.3.2}$$

The existence of such a solution is assured by Theorem 10.1, since a is bounded by assumption (10.1.6), $|T_n(u_n)| \leq n$ and the second term is bounded by $\nu n^p|\nabla u_n|^2$, due to (10.1.7). Choosing u_n^- as a test function, it is easy to see that u_n is positive, since $f \geq 0$. Now we use u_n as a test function. By (10.1.6) and (10.1.7), dropping positive terms, we have

$$\mu \int_\Omega |\nabla u_n|^2 \leq \int_\Omega f_n u_n\,.$$

It is easy to deduce that the sequence u_n is bounded in $H_0^1(\Omega)$. Moreover, the use of $G_k(u_n)$ as test function yields, dropping again positive terms,

$$\alpha \int_\Omega |\nabla G_k(u_n)|^2 \leq \int_\Omega f G_k(u_n)\,.$$

By Remark 6.8 we infer that the sequence u_n is bounded in $L^\infty(\Omega)$.

Step II: With the same technique as in the previous section, one can prove that u_n converges in $H_0^1(\Omega)$ to some function $u \in H_0^1(\Omega) \cap L^\infty(\Omega)$. Indeed, up to a subsequence, there exists $u \in H_0^1(\Omega)$ such that $\nabla u_n \to \nabla u$ weakly in $L^2(\Omega)$. Let us consider as a test function $v_n = \psi(u_n - u)$, where $\psi(t) = (e^{\lambda|t|} - 1)\mathrm{sgn}(t)$. We can write

$$\begin{aligned}
&\int_\Omega [a(x) + |T_n(u_n)|^q](\nabla u_n - \nabla u) \cdot (\nabla u_n - \nabla u)\psi'(u_n - u) \\
&\quad + \int_\Omega b(x) T_n(u_n)|T_n(u_n)|^{p-1}|\nabla u_n|^2 v_n \\
&= -\frac{1}{n}\int_\Omega u_n v_n - \int_\Omega [a(x) + |T_n(u_n)|^q]\nabla u \cdot (\nabla u_n - \nabla u)\psi'(u_n - u) + \int_\Omega f v_n .
\end{aligned}$$

By assumptions (10.1.6) and (10.1.7), one gets

$$\begin{aligned}
\alpha \int_\Omega |\nabla u_n - \nabla u|^2 \psi'(u_n - u) &\\
\le - \int_\Omega &[a(x) + |T_n(u_n)|^q]\nabla u \cdot \nabla(u_n - u)\psi'(u_n - u) \\
&- \frac{1}{n}\int_\Omega u_n v_n + \nu C \int_\Omega |\nabla u_n|^2 v_n + \int_\Omega f v_n ,
\end{aligned}$$

as $\|u_n\|_{L^\infty(\Omega)}$ is bounded. Now, $|\nabla u_n|^2 = |(\nabla u_n - \nabla u) + \nabla u|^2 \le 2|\nabla u_n - \nabla u|^2 + 2|\nabla u|^2$: this implies that

$$\begin{aligned}
&\int_\Omega |\nabla u_n - \nabla u|^2[\alpha\psi'(u_n - u) - 2\nu C|\psi(u_n - u)|] \\
&\quad \le 2\nu C \int_\Omega |\nabla u|^2 v_n - \int_\Omega [a(x) + |T_n(u_n)|^q]\nabla u \cdot (\nabla u_n - \nabla u)\psi'(u_n - u) \\
&\quad - \frac{1}{n}\int_\Omega u_n v_n + \int_\Omega f v_n .
\end{aligned}$$

Observe that, if $\lambda > \frac{2\gamma}{\alpha}$, then $\alpha\psi'(u_n - u) - 2\gamma|\psi(u_n - u)| \ge 1$. With this choice, using that $\|\psi'(u_n - u)\|_{L^\infty(\Omega)}$ and $\|u_n\|_{L^\infty(\Omega)}$ are bounded, one has

$$\begin{aligned}
\|\nabla u_n - \nabla u\|^2_{L^2(\Omega)} \le C \int_\Omega \nabla u \cdot (\nabla u_n - \nabla u) - \frac{1}{n}\int_\Omega u_n v_n \\
+ 2\nu C \int_\Omega |\nabla u|^2 v_n + \int_\Omega f v_n .
\end{aligned} \tag{10.3.3}$$

We now want to prove that the right-hand side of the previous inequality goes to 0. It is clear that $v_n \to 0$ a.e. in Ω and that the sequence v_n is bounded since ψ is continuous and u_n, u are bounded in $L^\infty(\Omega)$. Lebesgue's theorem implies that $\int_\Omega |\nabla u|^2 v_n \to 0$ and $\int_\Omega f v_n \to 0$. Using Hölder's inequality with exponent 2, one can prove that $\int_\Omega u_n v_n \to 0$. Since $\nabla u_n \to \nabla u$ weakly in $L^2(\Omega)$, the first term goes to 0. Therefore (10.3.3) implies that $\|\nabla u_n - \nabla u\|_{L^2(\Omega)} \to 0$.

Step III: It is now easy to pass to the limit in (10.3.2) to get a $H_0^1(\Omega) \cap L^\infty(\Omega)$ solution to

$$\begin{cases} -\mathrm{div}([a(x) + u^q]\nabla u) + b(x)u^p|\nabla u|^2 = f, & \text{in } \Omega, \\ u = 0, & \text{on } \partial\Omega. \end{cases} \qquad \square$$

In the following result we will get weak positive $H_0^1(\Omega)$ solutions to problem (10.1.5), in the sense that$b(x)u^p|\nabla u|^2 \in L^1(\Omega)$, and

$$\int_\Omega [a(x) + u^q]\nabla u \cdot \nabla\varphi + \int_\Omega b(x)u^p|\nabla u|^2\varphi = \int_\Omega f\varphi, \tag{10.3.4}$$

$$\text{for every} \quad \varphi \in H_0^1(\Omega) \cap L^\infty(\Omega). \tag{10.3.5}$$

Theorem 10.6. *Let Ω be a bounded open subset of $\mathbb{R}^N$, $N \geq 3$. Assume* (10.1.6) *and* (10.1.7). *Let f be a positive function belonging to $L^m(\Omega)$, $m \leq \frac{N}{2}$. Then there exists a solution u in the sense of* (10.3.5) *such that*
(1) *if $m = 1$, $p \geq 2q$, then u belongs to $L^{(p+2)\frac{N}{N-2}}(\Omega)$;*
(2) *if $\frac{2(q+1)N}{2N+p(N-2)+4q} \leq m \leq \frac{N}{2}$, $2q \geq p \geq q-1$, then u belongs to $L^{(p+2)m^{**}}(\Omega)$;*
(3) *if $\frac{2N}{N+2} \leq m \leq \frac{N}{2}$, $q \geq 1$, $2p \geq q-1 \geq p$, then u belongs to $L^{(q+1)m^{**}}(\Omega)$.*

In the case where the summability of f is lower, we will show the existence of a weak positive $H_0^1(\Omega)$ solution u to (10.1.5) in the sense that $b(x)u^p|\nabla u|^2 \in L^1(\Omega)$, and

$$\int_\Omega [a(x) + u^q]\nabla u \cdot \nabla\varphi + \int_\Omega b(x)u^p|\nabla u|^2\varphi = \int_\Omega f\varphi, \quad \text{for every} \quad \varphi \in W_0^{1,\infty}(\Omega). \tag{10.3.6}$$

Theorem 10.7. *Let Ω be a bounded open subset of $\mathbb{R}^N$, $N \geq 3$. Assume* (10.1.6) *and* (10.1.7), *$2q \geq p$. Let f be a positive function belonging to $L^m(\Omega)$, with $m \leq \frac{N}{2}$. Then there exists a weak solution in the sense of* (10.3.6) *such that*
(1) *if $1 \leq m \leq \frac{2(q+1)N}{2N+p(N-2)+4q}$, $p \geq q-1$, then $u \in L^{(p+2)m^{**}}(\Omega)$;*
(2) *if $\max[1, \frac{(2q-p)N}{2(2q-p)+(q+1)N}] \leq m \leq \frac{2N}{N+2}$, $q-1 \geq p$, then $u \in L^{(q+1)m^{**}}(\Omega)$.*

For the sake of simplicity we will present the above results in the case $q = p - 1$:

Theorem 10.8. *Let Ω be a bounded open subset of $\mathbb{R}^N$, $N \geq 3$. Assume* (10.1.6) *and* (10.1.7). *Let f be a positive function belonging to $L^m(\Omega)$, with $m \leq \frac{N}{2}$. Then there exists $0 \leq u \in H_0^1(\Omega)$, weak solution of* (10.1.5) *in the sense of* (10.3.5) *such that*
(1) *if $m = 1$, $p \leq 2$, then u belongs to $L^{(p+2)\frac{N}{N-2}}(\Omega)$;*
(2) *if $\frac{2pN}{2N+p(N-2)+4q} \leq m \leq \frac{N}{2}$, $p \geq 2$, then u belongs to $L^{(p+2)m^{**}}(\Omega)$.*

Theorem 10.9. *Let Ω be a bounded open subset of $\mathbb{R}^N$, $N \geq 3$. Assume* (10.1.6), (10.1.7) *and $p \geq 2$. Let f be a positive function belonging to $L^m(\Omega)$, with $m \leq \frac{N}{2}$. If $1 \leq m \leq \frac{2pN}{2N+p(N-2)+4(p-1)}$, then there exists $0 \leq u \in H_0^1(\Omega) \cap L^{(p+2)m^{**}}(\Omega)$ weak solution of* (10.1.5) *in the sense of* (10.3.6).

We will work by approximation on the following sequence of problems:

$$\begin{cases} -\operatorname{div}([a(x) + u_n^{p-1}]\nabla u_n) + b(x)u_n^p|\nabla u_n|^2 = f_n, & \text{in } \Omega, \\ u_n = 0, & \text{on } \partial\Omega, \end{cases} \tag{10.3.7}$$

where $f_n = T_n(f)$. These problems are well posed due to Theorem 10.5. In the following lemma we prove that the sequence u_n is bounded in $H_0^1(\Omega)$.

Lemma 10.10. *Let $f \in L^1(\Omega)$.*
(1) *There exists $R > 0$ such that*

$$\|u_n\|^2_{H_0^1(\Omega)} \leq R\|f\|_{L^1(\Omega)}; \tag{10.3.8}$$

(2) *if $A_k = \{u_n \geq k\}$, then*

$$\int_{A_k} b(x)u_n^p|\nabla u_n|^2 \leq \int_{A_k} f. \tag{10.3.9}$$

Moreover, there exists $u \in H_0^1(\Omega)$ such that $u_n \to u$ in $H_0^1(\Omega)$ up to a subsequence.

Proof. The use of $T_j(u_n)$ as a test function in (10.3.7) implies

$$\int_\Omega [a(x) + u_n^{p-1}]\nabla u_n \cdot \nabla T_j(u_n) + \int_\Omega b(x)u_n^p|\nabla u_n|^2 T_j(u_n) \leq \int_\Omega f_n T_j(u_n).$$

By assumptions (10.1.6) and (10.1.7) one has

$$\alpha \int_\Omega |\nabla T_j(u_n)|^2 + \mu\, j^{p+1} \int_{\{j<u_n\}} |\nabla u_n|^2 \leq j \int_\Omega f.$$

We deduce that, for every $j \in \mathbb{N}$,

$$\int_\Omega |\nabla u_n|^2 = \int_{\{0\leq u_n<j\}} |\nabla u_n|^2 + \int_{\{j\leq u_n\}} |\nabla u_n|^2 \leq \left[\frac{j}{\alpha} + \frac{1}{\mu\, j^p}\right] \|f\|_{L^1(\Omega)}$$

which gives (10.3.8). Therefore there exist $u \in H_0^1(\Omega)$ and a subsequence (still denoted by u_n) such that we have that u_n converges weakly to u in $H_0^1(\Omega)$.

The use of $T_j(u_n - T_k(u_n))$ as a test function in (10.3.7) gives, dropping positive terms,

$$\int_{\{|u_n - T_k(u_n)| \le j\} \cap A_k} b(x) u_n^p |\nabla u_n|^2 \le j \int_{A_k} f .$$

The limit as $j \to \infty$ implies (10.3.9). □

Lemma 10.11. *The sequence $\nabla u_n(x)$ converges a.e. in Ω to $\nabla u(x)$, where u is the function found in Lemma 10.10.*

Proof. We use $T_h(u_n - T_k(u))$ as a test function in (10.3.7): then

$$\int_\Omega [a(x) + u_n^{p-1}] \nabla u_n \cdot \nabla T_h(u_n - T_k(u)) \le 2h \|f\|_{L^1(\Omega)} .$$

This gives, by (10.3.9),

$$\begin{aligned} \int_\Omega \Big\{ [a(x) + u_n^{p-1}] \nabla u_n - [a(x) + u_n^{p-1}] \nabla T_k(u) \Big\} \cdot \nabla T_h(u_n - T_k(u)) \\ \le 2h \|f\|_{L^1(\Omega)} - \int_\Omega [a(x) + u_n^{p-1}] \nabla T_k(u) \cdot \nabla T_h(u_n - T_k(u)) . \end{aligned}$$

Now, $\nabla T_h(u_n - T_k(u))$ is not zero on $\{-h + T_k(u) \le u_n \le h + T_k(u)\}$. Therefore $\int_\Omega [a(x) + u_n^{p-1}] \nabla T_k(u) \cdot \nabla T_h(u_n - T_k(u)) \to 0$, since $T_h(u_n - T_k(u)) \to T_h(u - T_k(u))$ weakly in $(L^2(\Omega))^N$ and the sequence u_n is bounded on this set. Thus it follows from (10.1.6) that

$$\alpha \limsup_{n \to \infty} \int_\Omega |\nabla T_h(u_n - T_k(u))|^2 \le 2h \|f\|_{L^1(\Omega)} . \tag{10.3.10}$$

Let now r be such that $1 < r < 2$ and R as in Lemma 10.10. It is clear that

$$\begin{aligned} \int_\Omega |\nabla(u_n - u)|^r = \int_\Omega |\nabla T_h(u_n - T_k(u))|^r + \int_{\{|u_n - u| \le h,\, u > k\}} |\nabla(u_n - u)|^r \\ + \int_{\{|u_n - u| > h\}} |\nabla(u_n - u)|^r . \end{aligned}$$

By Hölder's inequality with exponent $\frac{2}{r}$ on the last two terms and estimate (10.3.8) we deduce that

$$\begin{aligned} \int_\Omega |\nabla(u_n - u)|^r \le \int_\Omega |\nabla T_h(u_n - T_k(u))|^r \\ + 2^{r-1} R^r \|f\|_{L^1(\Omega)}^r \operatorname{meas}(\{u > k\})^{1 - \frac{r}{2}} \\ + 2^{r-1} R^r \|f\|_{L^1(\Omega)}^r \operatorname{meas}(\{|u_n - u| > h\})^{1 - \frac{r}{2}} . \end{aligned}$$

Thus, for every $h > 0$ and $k > 0$, we deduce from (10.3.10) and the $L^2(\Omega)$ convergence of u_n to u, that

$$\limsup_{n\to\infty} \int_\Omega |\nabla(u_n - u)|^r \le \left[h\,\frac{2}{\alpha}\int_\Omega f\right]^{\frac{r}{2}} \operatorname{meas}(\Omega)^{1-\frac{r}{2}} + 2^{r-1}R^r \operatorname{meas}(\{u > k\})^{1-\frac{r}{2}}.$$

The limit as $h \to 0$ and then $k \to +\infty$ gives

$$\int_\Omega |\nabla(u_n - u)|^r \to 0, \quad \forall\, r < 2. \tag{10.3.11}$$

Then (up to subsequences) $\nabla u_n(x)$ converges a.e. in Ω to $\nabla u(x)$. □

Fatou's lemma implies the following corollary.

Corollary 10.12. *Let u be the function found in Lemma 10.10. Then*

$$0 \le \int_\Omega b(x) u^p |\nabla u|^2 \le \int_\Omega f.$$

Proof. It is sufficient to pass to the limit in (10.3.9) written for $k = 0$ using the previous lemma. □

Remark 10.13. We point out that in Lemma 10.10 and in Lemma 10.11 we only require f to belong to $L^1(\Omega)$.

Lemma 10.14. *Under the assumptions of Theorem 10.8, $[a(x) + u_n^{p-1}]\nabla u_n$ converges weakly in $(L^2(\Omega))^N$ to $[a(x) + u^{p-1}]\nabla u$ and a.e. in Ω.*

Proof. By the previous lemmata it is sufficient to prove that the sequence $[a(x) + u_n^{p-1}]\nabla u_n$ is bounded in $L^2(\Omega)$.

Step I: Assume $m = 1$ and $p \le 2$. Estimate (10.3.9) and assumption (10.1.7) imply that

$$\mu \int_\Omega u_n^p |\nabla u_n|^2 \le \int_\Omega f. \tag{10.3.12}$$

Since $p \le 2$, by (10.1.6), (10.3.8) and (10.3.12) we get

$$\begin{aligned}\int_\Omega [a(x) + u_n^{p-1}]^2 |\nabla u_n|^2 &\le \int_{\{1\le u_n\}} 2[\beta^2 + u_n^p]|\nabla u_n|^2 + \int_{\{u_n<1\}} 2[\beta^2 + 1]|\nabla u_n|^2 \\ &\le 2(\beta^2+1)\int_\Omega |\nabla u_n|^2 + 2\frac{\|f\|_{L^1(\Omega)}}{\mu} \\ &\le 2(\beta^2+1)R\|f\|_{L^1(\Omega)} + 2\frac{\|f\|_{L^1(\Omega)}}{\mu}.\end{aligned}$$

The a.e. convergence of u_n and of ∇u_n imply the result.

Step II: Assume $p \geq 2$ and $1 \leq m \leq \frac{2pN}{2N+p(N-2)+4(p-1)}$. We first prove that u_n is bounded in $L^{(p+2)m^{**}}(\Omega)$. Let

$$r = \frac{(p+2)m^{**}}{2^*}.$$

We use $(\varepsilon + u_n)^{2r-2-p} - \varepsilon^{2r-2-p}$, $0 < \varepsilon < 1$, as a test function in (10.3.7). Dropping positive terms and using (10.1.7) we get

$$\begin{aligned}\mu \int_\Omega u_n^{2(r-1)} |\nabla u_n|^2 &\leq C \int_\Omega |f|(\varepsilon + u_n)^{2r-2-p} \\ &\leq C\|f\|_{L^1(\Omega)} + C\|f\|_{L^m(\Omega)} \left[\int_\Omega |u_n|^{(2r-2-p)m'}\right]^{\frac{1}{m'}}.\end{aligned} \tag{10.3.13}$$

By Sobolev inequality on the left-hand side, we deduce that the sequence u_n is bounded in $L^{(p+2)m^{**}}(\Omega)$, as $r2^* = (2r-2-p)m' = (p+2)m^{**} \geq 0$ and $\frac{2}{2^*} > \frac{1}{m'}$. As a consequence of (10.3.13), also the sequence $\int_\Omega u_n^{2(r-1)} |\nabla u_n|^2$ is bounded. Since $p \leq r$ this implies that

$$\begin{aligned}\int_\Omega [a(x) + u_n^{p-1}]^2 |\nabla u_n|^2 &\leq 2\beta^2 \int_\Omega |\nabla u_n|^2 + 2\int_\Omega u_n^{2(p-1)} |\nabla u_n|^2 \\ &\leq 2\beta^2 R\|f\|_{L^1(\Omega)} + C.\end{aligned}$$

□

Proof of Theorem 10.8. We will first prove that

$$\int_\Omega [a(x) + u^{p-1}]\nabla u \cdot \nabla\varphi + \int_\Omega b(x)u^p|\nabla u|^2\varphi = \int_\Omega f\varphi$$

for every $0 \leq \varphi \in H_0^1(\Omega) \cap L^\infty(\Omega)$. Then we will treat the general case.

Step I: We have

$$\int_\Omega [a(x) + u_n^{p-1}]\nabla u_n \cdot \nabla\varphi + \int_\Omega b(x)u_n^p|\nabla u_n|^2\varphi = \int_\Omega f_n\varphi,$$

for every $0 \leq \varphi \in H_0^1(\Omega) \cap L^\infty(\Omega)$. Lemma 10.14 and Fatou's lemma yield

$$\int_\Omega [a(x) + u^{p-1}]\nabla u \cdot \nabla\varphi + \int_\Omega b(x)u^p|\nabla u|^2\varphi \leq \int_\Omega f\varphi. \tag{10.3.14}$$

Step II: Let $0 \leq \varphi \in H_0^1(\Omega)$. We define $\gamma(s) = s^p$, $H(t) = \int_0^t \gamma(s)\,ds$ and use

$$e^{-\frac{\nu}{\alpha}H(u_n)} e^{\frac{\nu}{\alpha}H(T_k(u))}\varphi$$

as a test function. We obtain

$$\begin{aligned}
&\int_\Omega [a(x) + u_n^{p-1}]\nabla u_n \cdot \nabla\varphi e^{-\frac{\nu}{\alpha}H(u_n)} e^{\frac{\nu}{\alpha}H(T_k(u))} \\
&\qquad + \frac{\nu}{\alpha}\int_\Omega [a(x) + u_n^{p-1}]\nabla u_n \cdot \nabla T_k(u)\gamma(T_k(u))e^{-\frac{\nu}{\alpha}H(u_n)} e^{\frac{\nu}{\alpha}H(T_k(u))}\varphi \\
&\quad = \int_\Omega f_n e^{-\frac{\nu}{\alpha}H(u_n)} e^{\frac{\nu}{\alpha}H(T_k(u))}\varphi \\
&\qquad + \frac{\nu}{\alpha}\int_\Omega [a(x) + u_n^{p-1}]\nabla u_n \cdot \nabla u_n \gamma(u_n)e^{-\frac{\nu}{\alpha}H(u_n)} e^{\frac{\nu}{\alpha}H(T_k(u))}\varphi \\
&\qquad - \int_\Omega b(x)u_n^p|\nabla u_n|^2 e^{-\frac{\nu}{\alpha}H(u_n)} e^{\frac{\nu}{\alpha}H(T_k(u))}\varphi \ge 0\,.
\end{aligned}$$

The limit $n \to \infty$, Fatou's lemma and Lemma 10.14 yield

$$\begin{aligned}
&\int_\Omega [a(x) + u^{p-1}]\nabla u \cdot \nabla\varphi\, e^{-\frac{\nu}{\alpha}H(u)}\, e^{\frac{\nu}{\alpha}H(T_k(u))} \\
&\qquad + \frac{\nu}{\alpha}\int_\Omega [a(x) + u^q]\nabla u \cdot \nabla T_k(u)\gamma(T_k(u))\, e^{-\frac{\nu}{\alpha}H(u)} e^{\frac{\nu}{\alpha}H(T_k(u))}\varphi \\
&\ge \int_\Omega f e^{-\frac{\nu}{\alpha}H(u)} e^{\frac{\nu}{\alpha}H(T_k(u))}\varphi \frac{\nu}{\alpha}\int_\Omega [a(x) + u^{p-1}]\nabla u \cdot \nabla u\, \gamma(u)e^{-\frac{\nu}{\alpha}H(u)} e^{\frac{\nu}{\alpha}H(T_k(u))}\varphi \\
&\quad - \int_\Omega b(x)u^p|\nabla u|^2 e^{-\frac{\nu}{\alpha}H(u)} e^{\frac{\nu}{\alpha}H(T_k(u))}\varphi\,.
\end{aligned}$$

This implies that

$$\begin{aligned}
&\int_\Omega [a(x) + u^{p-1}]\nabla u \cdot \nabla\varphi\, e^{-\frac{\nu}{\alpha}H(u)} e^{\frac{\nu}{\alpha}H(T_k(u))} \\
&\qquad \ge \int_\Omega f e^{-\frac{\nu}{\alpha}H(u)} e^{\frac{\nu}{\alpha}H(T_k(u))}\varphi - \int_\Omega b(x)u^p|\nabla u|^2 e^{-\frac{\nu}{\alpha}H(u)} e^{\frac{\nu}{\alpha}H(T_k(u))}\varphi\,.
\end{aligned}$$

In order to use Lebesgue's theorem (as $k \to +\infty$) in the previous inequality, note that $\frac{e^{\frac{\nu}{\alpha}H(T_k(u))}}{e^{\frac{\nu}{\alpha}H(u)}} \le 1$. Then

$$\int_\Omega [a(x) + u^{p-1}]\nabla u \cdot \nabla\varphi \ge \int_\Omega f\varphi - \int_\Omega b(x)u^p|\nabla u|^2\varphi\,. \tag{10.3.15}$$

Lemma 10.14 implies that $u^{p-1}|\nabla u| \in L^2(\Omega)$; inequalities (10.3.14) and (10.3.15) imply, for every $0 \le \varphi \in H_0^1(\Omega) \cap L^\infty(\Omega)$,

$$\int_\Omega [a(x) + u^{p-1}]\nabla u \cdot \nabla\varphi + \int_\Omega b(x)u^p|\nabla u|^2\varphi = \int_\Omega f\varphi\,.$$

Step III: Every positive $\varphi \in H_0^1(\Omega) \cap L^\infty(\Omega)$ can be written as $\varphi = \varphi^+ - \varphi^-$. Therefore we can prove the existence of a solution u of the Dirichlet problem (10.3.4). □

Proof of Theorem 10.9. We point out only the differences with respect to the proof of Theorem 10.8. First of all we observe that, since $p \geq 2$, by Corollary 10.12 and Sobolev's inequality, u_n is bounded in $L^{p-2}(\Omega)$. We are going to prove that the sequence $u_n^{p-1}|\nabla u_n|$ is compact in $L^1(\Omega)$. By (10.3.9) and Hölder's inequality we have

$$\int_{A_k} u_n^{p-1}|\nabla u_n| \leq \left[\int_{A_k} u_n^{p-2}\right]^{\frac{1}{2}} \left[\int_{A_k} u_n^p|\nabla u_n|^2\right]^{\frac{1}{2}} \leq C\left[\int_{A_k} f\right]^{\frac{1}{2}}.$$

Then, for every measurable subset E, the inequalities

$$\begin{aligned}\int_E u_n^{p-1}|\nabla u_n| &\leq \int_{\{k\leq u_n\}} u_n^{p-1}|\nabla u_n| + k^{p-1}\int_E |\nabla u_n| \\ &\leq C\left[\int_{A_k} f\right]^{\frac{1}{2}} + k^q R^{\frac{1}{2}} \operatorname{meas}(E)^{\frac{1}{2}} \|f\|_{L^1(\Omega)}^{\frac{1}{2}}\end{aligned}$$

imply

$$\limsup_{\operatorname{meas}(E)\to 0} \int_E u_n^{p-1}|\nabla u_n| \leq C\left[\int_{A_k} f\right]^{\frac{1}{2}}.$$

Thus Vitali's theorem yields that $u_n^{p-1}|\nabla u_n|$ strongly converges in $L^1(\Omega)$ to $u^{p-1}|\nabla u|$. To pass to the limit in (10.3.7), we repeat the proof of Theorem 10.8, obtaining two inequalities; the first one can be obtained exactly as before, while for the second one we have to modify the proof, since we no longer have (10.14). For $j > 0$, define

$$R_j(s) = \begin{cases} 1 & \text{if } s \leq j, \\ j+1-s & \text{if } j \leq s < j+1, \\ 0 & \text{if } s > j+1, \end{cases}$$

and choose $e^{-\frac{\nu}{\alpha}H(u_n)}e^{\frac{\nu}{\alpha}H(T_k(u))}R_j(u_n)\,\varphi$ as a test function in (10.3.7). We conclude the proof, as in the Theorem 10.8 (proof of (10.3.15)), letting first k tend to infinity, and then j tend to infinity, observing that $R_j(s)$ tends to 1. □

11 Problems with low summable sources

11.1 Introduction

In Chapters 5 and 6, we focused our attention on the existence of solutions to the Leray–Lions problem

$$\begin{cases} -\operatorname{div}(a(x,u,\nabla u)) = f\,, & \text{in} \quad \Omega \\ u = 0\,, & \text{on} \quad \partial\Omega\,, \end{cases} \tag{11.1.1}$$

assuming that the source f belongs to $L^m(\Omega)$ with $m \geq \frac{2N}{N+2}$ and under the following assumptions on a Carathéodory map $a : \Omega \times \mathbb{R} \times \mathbb{R}^N \to \mathbb{R}^N$:

(1) there exists $\beta > 0$ such that $|a(x,s,\xi)| \leq \beta[|s| + |\xi|]$;
(2) there exists $\alpha > 0$ such that $a(x,s,\xi) \cdot \xi \geq \alpha|\xi|^2, \quad \forall\, \xi \in \mathbb{R}^N$;
(3) $[a(x,s,\xi) - a(x,s,\eta)] \cdot [\xi - \eta] > 0$ if $\xi \neq \eta$.

Moreover Ω is an open bounded subset of $\mathbb{R}^N$, $N \geq 3$.

In this chapter we study the existence and the summability of the solutions in the case where f belongs to $L^m(\Omega)$ with $1 \leq m < \frac{2N}{N+2}$. Note that we cannot use the Leray–Lions theorem (Theorem 5.1), since the source does not belong to $H^{-1}(\Omega)$. We will prove the existence of distributional solutions u, that is u will be a function in $W_0^{1,1}(\Omega)$, at least, such that

$$\int_\Omega a(x,u,\nabla u) \cdot \nabla\varphi = \int_\Omega f\varphi\,, \quad \forall\, \varphi \in C_0^\infty(\Omega)\,.$$

Although u does not belong to $H_0^1(\Omega)$, we will show that, for every $k > 0$,

$$\int_\Omega |\nabla T_k(u)|^2 \leq C_0\, k,$$

which is fundamental for the uniqueness of solutions, as we will see in Chapter 12.

For the reader's convenience we summarize here the main results of existence of solutions u in function of the summability of the source f:

$$f \in L^1(\Omega) \text{ or } f \text{ is a bounded measure} \Rightarrow u \in M^{\frac{N}{N-2}}(\Omega),\, |\nabla u| \in M^{\frac{N}{N-1}}(\Omega)\,;$$
$$f \in L^m(\Omega),\ 1 < m \leq \frac{2N}{N+2} \Rightarrow u \in W_0^{1,m^*}(\Omega)\,.$$

The last result can be seen as a nonlinear Calderon–Zygmund theory for elliptic problems with infinite energy solutions.

We will follow the proofs of [12, 13]. The reader can find further references there; for recent developments see [38].

In the linear case, the existence results (if f is a bounded measure or if f belongs to $L^m(\Omega)$, $1 \le m \le \frac{2N}{N+2}$) are due to G. Stampacchia. They can be seen as a Calderon–Zygmund theory for linear operators with discontinuous coefficients.

We observe that general results can be proved in the case where a satisfies the coercivity assumption

$$a(x,s,\xi)\cdot\xi \ge \alpha|\xi|^p\,, \quad \forall\,\xi \in \mathbb{R}^N$$

(see Remark 11.18).

Moreover, we will study the regularizing effects of a lower order term. More precisely, we will consider the problem

$$\begin{cases} -\mathrm{div}(a(x,u,\nabla u)) + |u|^{p-1}u = f\,, & \text{in } \Omega\,,\\ u = 0\,, & \text{on } \partial\Omega\,, \end{cases}$$

for a source $f \in L^m(\Omega)$. It is not difficult to prove that $|u|^p \in L^m(\Omega)$. We will prove that

$$f \in L^m(\Omega),\ m > 1, p \ge \frac{1}{m-1} \Rightarrow u \in H_0^1(\Omega);$$
$$f \in L^m(\Omega),\ m > 1, p < \frac{1}{m-1} \Rightarrow u \in W_0^{1,q}(\Omega)\,, \forall\, q < \frac{2pm}{1+pm};$$
$$f \in L^1(\Omega) \Rightarrow u \in W_0^{1,q}(\Omega)\,, \quad \forall\, q < \frac{2p}{1+p}\,,$$

following the proofs in [16]. Note that the study by Brezis and Strauss of semilinear equations like

$$\begin{cases} -\mathrm{div}(M(x)\nabla u) + |u|^{p-1}u = f\,, & \text{in } \Omega\,,\\ u = 0\,, & \text{on } \partial\Omega\,, \end{cases}$$

was the bridge between the linear case and the nonlinear case.

In the last section we will study the regularity of minimizers for integral functionals of the form

$$J(v) = \int_\Omega j(x,\nabla v) - \int_\Omega f v$$

where $j : \Omega \times \mathbb{R}^N \to \mathbb{R}$ is a Carathéodory function, convex in the last variable, such that

$$j(x,\xi) \ge \alpha|\xi|^2$$

for some positive α and f belongs $L^1(\Omega)$. Due to the summability of f, the nonexistence of minima led to the definition of T-minima, in analogy with the definition of entropy solutions to elliptic problems (see [5] and [33]).

11.2 A priori estimates

We will work by approximation to prove the existence of distributional solutions and entropy solutions to problem (11.1.1). We consider

$$\begin{cases} -\operatorname{div}(a(x,u_n,\nabla u_n)) = f_n\,, & \text{in } \Omega\,, \\ u_n = 0\,, & \text{on } \partial\Omega\,, \end{cases} \tag{11.2.1}$$

where f_n is a sequence of functions in $H^{-1}(\Omega)\cap L^\infty(\Omega)$ such that $f_n \to f$ in $L^m(\Omega)$, $\|f_n\|_{L^m(\Omega)} \le \|f\|_{L^m(\Omega)}$ and $|f_n(x)| \le |f(x)|$ a.e. in Ω (for example $f_n = T_n(f)$). The existence of solutions u_n, for every n, follows from Theorem 5.1; moreover u_n belongs to $H_0^1(\Omega)\cap L^\infty(\Omega)$ due to Theorem 6.6. Our strategy is as follows. We get the following estimates: the sequence u_n is bounded in $W_0^{1,m^*}(\Omega)$ when $m > 1$ and in $W_0^{1,q}(\Omega)$, $q < \frac{N}{N-1}$, when $m = 1$. This allows us to get a subsequence converging to some function u. We prove that $\nabla u_n \to \nabla u$ a.e. in Ω. In this way we can pass to the limit in (11.2.1) and prove that u is a solution to problem (11.1.1).

Lemma 11.1. *Let $f \in L^1(\Omega)$. Then the sequence of solutions u_n to problems* (11.2.1) *is bounded in $W_0^{1,q}(\Omega)$ with $q < \frac{N}{N-1}$.*

Proof. We consider $v_n = [(1+|u_n|)^{2\lambda-1} - 1]\operatorname{sgn}(u_n)$ as a test function in (11.2.1). Let $\lambda < 1/2$: in this way $|v_n| \le 1$. As for the right-hand side we have

$$\int_\Omega f_n v_n \le \int_\Omega |f_n| \le \|f\|_{L^1(\Omega)}\,. \tag{11.2.2}$$

Using the ellipticity of a in the left-hand side we have

$$\begin{aligned} \|f\|_{L^1(\Omega)} \ge \int_\Omega a(x,u_n,\nabla u_n)\cdot\nabla v_n &\ge \alpha\int_\Omega |\nabla u_n|^2(1-2\lambda)(1+|u_n|)^{2\lambda-2} \\ &= (1-2\lambda)\alpha\int_\Omega \left|\frac{\nabla[(1+|u_n|)^\lambda]}{\lambda}\right|^2. \end{aligned}$$

The above estimates imply

$$\int_\Omega \frac{|\nabla u_n|^2}{(1+|u_n|)^{2(1-\lambda)}} \le \frac{\|f\|_{L^1(\Omega)}}{\alpha(1-2\lambda)}\,. \tag{11.2.3}$$

On the other hand

$$\int_\Omega |\nabla u_n|^q = \int_\Omega \frac{|\nabla u_n|^q}{(1+|u_n|)^{2(1-\lambda)\frac{q}{2}}}(1+|u_n|)^{2(1-\lambda)\frac{q}{2}}\,.$$

Using Sobolev's inequality on the left-hand side and Hölder's inequality with exponent $\frac{2}{q}$ in the right one, we obtain

$$S^q \left(\int_\Omega |u_n|^{q^*} \right)^{\frac{q}{q^*}} \le \int_\Omega |\nabla u_n|^q \le \left(\int_\Omega \frac{|\nabla u_n|^2}{(1+|u_n|)^{2(1-\lambda)}} \right)^{\frac{q}{2}} \left(\int_\Omega (1+|u_n|)^{\frac{(1-\lambda)2q}{2-q}} \right)^{1-\frac{q}{2}} .$$

Inequality (11.2.3) implies

$$S^q \left(\int_\Omega |u_n|^{q^*} \right)^{\frac{q}{q^*}} \le \int_\Omega |\nabla u_n|^q \le C + C \left(\int_\Omega |u_n|^{\frac{(1-\lambda)2q}{2-q}} \right)^{1-\frac{q}{2}} \tag{11.2.4}$$

where C denotes a constant depending on $q, \lambda, \alpha, \|f\|_{L^1(\Omega)}$, independent on n. Let us choose λ such that $\frac{(1-\lambda)2q}{2-q} = q^*$, that is, $\lambda = \frac{q(N-2)}{2(N-q)}$. Since $\lambda < \frac{1}{2}$, this implies $q < \frac{N}{N-1}$. In this way

$$S^q \left(\int_\Omega |u_n|^{q^*} \right)^{\frac{q}{q^*}} \le C + C \left(\int_\Omega |u_n|^{q^*} \right)^{1-\frac{q}{2}}$$

and so the sequence $\int_\Omega |u_n|^{q^*}$ is bounded. Hence the right-hand side of (11.2.4) is uniformly bounded and $\int_\Omega |\nabla u_n|^q$ too. □

Lemma 11.2. *Let $f \in L^m(\Omega)$, $m > 1$. Then the sequence of the solutions u_n to problems* (11.2.1) *is bounded in $W_0^{1,m^*}(\Omega)$.*

Proof. We divide the proof into two steps.

Step I: We prove that the sequence u_n is bounded in $L^{m^{**}}(\Omega)$. Let us consider $v_n = [(1+|u_n|)^{2\lambda-1} - 1]\operatorname{sgn}(u_n)$ as a test function in (11.2.1); $\lambda > \frac{1}{2}$ will be defined later. For the right-hand side we have,

$$\int_\Omega f_n v_n \le \int_\Omega |f|[(1+|u_n|)^{2\lambda-1} - 1] \le \int_\Omega |f|[(1+|u_n|)^{2\lambda-1} + 1] .$$

By Hölder's inequality with exponent m we get

$$\int_\Omega f_n v_n \le \|f\|_{L^1(\Omega)} + \|f\|_{L^m(\Omega)} \left(\int_\Omega (1+|u_n|)^{(2\lambda-1)m'} \right)^{\frac{1}{m'}} .$$

For the left-hand side of (11.2.1), using the ellipticity of a and Sobolev's inequality, we have

$$\int_\Omega a(x,u_n,\nabla u_n)\cdot\nabla v_n \ge \alpha\int_\Omega |\nabla u_n|^2(2\lambda-1)(1+|u_n|)^{2\lambda-2}$$
$$= (2\lambda-1)\alpha\int_\Omega\left|\frac{\nabla[(1+|u_n|)^\lambda]}{\lambda}\right|^2$$
$$\ge \frac{(2\lambda-1)\alpha S}{\lambda^2}\left(\int_\Omega[(1+|u_n|)^\lambda]^{2^*}\right)^{\frac{2}{2^*}}.$$

Summarizing

$$\|f\|_{L^1(\Omega)}+\|f\|_{L^m(\Omega)}\left(\int_\Omega(1+|u_n|)^{(2\lambda-1)m'}\right)^{\frac{1}{m'}}$$
$$\ge(2\lambda-1)\alpha\int_\Omega\frac{|\nabla u_n|^2}{(1+|u_n|)^{2(1-\lambda)}}$$
$$\ge\frac{(2\lambda-1)\alpha S^2}{\lambda^2}\left(\int_\Omega[(1+|u_n|)^\lambda]^{2^*}\right)^{\frac{2}{2^*}}. \tag{11.2.5}$$

Now, let us fix λ such that $\lambda 2^*=(2\lambda-1)m'$, that is, $\lambda=\frac{m^{**}}{2^*}(>1/2)$. With this choice, we deduce from the previous estimate that

$$\frac{(2\lambda-1)\alpha S^2}{\lambda^2}\left(\int_\Omega|(1+|u_n|)^{\frac{m^{**}}{2^*}}|^{2^*}\right)^{\frac{2}{2^*}}$$
$$\le\|f\|_{L^1(\Omega)}+\|f\|_{L^m(\Omega)}\left(\int_\Omega|1+u_n|^{m^{**}}\right)^{\frac{1}{m'}}.$$

Since $\frac{2}{2^*}>\frac{1}{m'}$ this implies

$$\int_\Omega|u_n|^{m^{**}}\le\int_\Omega|1+u_n|^{m^{**}}\le C, \tag{11.2.6}$$

where C denotes a constant depending on $S, m, \alpha, \|f\|_{L^1(\Omega)}, \|f\|_{L^m(\Omega)}$.

Step II: Let us prove that the sequence u_n is bounded in $W_0^{1,m^*}(\Omega)$. We observe that estimates (11.2.5) and (11.2.6) imply that the sequence $\int_\Omega\frac{|\nabla u_n|^2}{(1+|u_n|)^{2(1-\lambda)}}$ is bounded. Assume that $\lambda<1$, that is, $1<m<\frac{2N}{N+2}$. Let $1\le q<2$ and write $\int_\Omega|\nabla u_n|^q$ as

$$\int_\Omega|\nabla u_n|^q=\int_\Omega\frac{|\nabla u_n|^q}{(1+|u_n|)^{2(1-\lambda)\frac{q}{2}}}(1+|u_n|)^{2(1-\lambda)\frac{q}{2}}.$$

By Hölder's inequality with exponent $\frac{2}{q}$ we get

$$\int_\Omega |\nabla u_n|^q \le \left(\int_\Omega \frac{|\nabla u_n|^2}{(1+|u_n|)^{2(1-\lambda)}}\right)^{\frac{q}{2}} \left(\int_\Omega (1+|u_n|)^{(1-\lambda)\frac{2q}{2-q}}\right)^{1-\frac{q}{2}}.$$

Let $q = m^*$; in this way $\frac{(1-\lambda)2q}{2-q} = m^{**}$. From (11.2.6) it follows that the sequence $\int_\Omega |\nabla u_n|^{m^*}$ is bounded. □

The following lemma will be useful for us:

Lemma 11.3. (1) *Let $f \in L^1(\Omega)$. Then, if u_n are the $H_0^1(\Omega)$ solutions to* (11.2.1), *there exists $u \in W_0^{1,q}(\Omega), q < \frac{N}{N-1}$, such that $\nabla u_n \to \nabla u$ a.e. in Ω, up to a subsequence.*
(2) *If $f \in L^m(\Omega), m > 1, u \in W_0^{1,m^*}(\Omega)$.*

Proof. The estimates of the previous lemmata allow us to get a subsequence, still denoted by u_n, weakly converging to some $u \in W_0^{1,q}(\Omega), q < \frac{N}{N-1}$. We note that if $f \in L^m(\Omega), m > 1$, then u belongs to $W_0^{1,m^*}(\Omega)$.

To prove the result, we will prove that for $\theta \in \left(0, \frac{q}{4}\right)$ one has

$$\lim_{n\to\infty} \int_\Omega \{[a(x,u_n,\nabla u_n) - a(x,u,\nabla u)] \cdot \nabla(u_n - u)\}^\theta = 0 \tag{11.2.7}$$

and then we will use Lemma 5.8. In the proof C denotes a constant independent of n (depending on $\beta, \theta, \text{meas}(\Omega)$).

We write Ω as the union of A_k and C_k, where

$$A_k := \{|u_n| \ge k\}, \quad C_k := \{|u_n| \le k\}.$$

Step I: Let us estimate

$$\int_{A_k} a(x,u_n,\nabla u_n) - a(x,u,\nabla u) \cdot \nabla(u_n - u)^\theta.$$

By the growth assumption on a we get

$$\begin{aligned}
&\int_{A_k} |[a(x,u_n,\nabla u_n) - a(x,u,\nabla u)] \cdot \nabla(u_n - u)|^\theta \\
&\le \int_{A_k} |a(x,u_n,\nabla u_n) \cdot \nabla(u_n - u)|^\theta + \int_{A_k} |a(x,u,\nabla u) \cdot \nabla(u_n - u)|^\theta \\
&\le 2\beta \int_{A_k} |u_n|^\theta |\nabla u_n|^\theta + 2\beta \int_{A_k} |u_n|^\theta |\nabla u|^\theta + 2\beta \int_{A_k} |u|^\theta |\nabla u_n|^\theta \\
&\quad + 2\beta \int_{A_k} |u|^\theta |\nabla u|^\theta + 2\beta \int_{A_k} |\nabla u_n|^{2\theta} + 2\beta \int_{A_k} |\nabla u|^{2\theta} + 2\beta \int_{A_k} |\nabla u|^\theta |\nabla u_n|^\theta.
\end{aligned}$$

The Cauchy–Schwarz inequality and the estimate on $\|u_n\|_{W_0^{1,q}(\Omega)}$ of the above lemmata give

$$\int_{A_k} |[a(x,u_n,\nabla u_n) - a(x,u_n,\nabla u)]\cdot\nabla(u_n-u)|^\theta \le C\,\mathrm{meas}(A_k)^{1/2}.$$

Step II: Let us now estimate

$$\int_{C_k} \{[a(x,u_n,\nabla u_n) - a(x,u,\nabla u)]\cdot\nabla(u_n-u)\}^\theta .$$

We observe that $|u| \le k$ on C_k (since $u_n \to u$ a.e. in Ω and $|u_n| \le k$ on C_k). Therefore one has

$$\begin{aligned}
&\int_{C_k} \{[a(x,u_n,\nabla u_n) - a(x,u_n,\nabla u)]\cdot\nabla(u_n-u)\}^\theta \\
&\quad = \int_{C_k} \{[a(x,u_n,\nabla u_n) - a(x,u,\nabla T_k(u))]\cdot\nabla(u_n-T_k(u))\}^\theta \\
&\quad \le \int_{\Omega} \{[a(x,u_n,\nabla u_n) - a(x,u,\nabla T_k(u))]\cdot\nabla(u_n-T_k(u))\}^\theta .
\end{aligned}$$

The last integral can be estimated as follows, if we set

$$V_j := \{|u_n - T_k(u)| \le j\}, \quad V_j' = \{|u_n - T_k(u)| > j\} :$$

$$\begin{aligned}
&\int_{C_k} \{[a(x,u_n,\nabla u_n) - a(x,u,\nabla u)]\cdot\nabla(u_n-u)\}^\theta \\
&\quad \le \int_{\Omega} \{[a(x,u_n,\nabla u_n) - a(x,u,\nabla T_k(u))]\cdot\nabla T_j(u_n-T_k(u))\}^\theta \\
&\qquad + \int_{V_j'} \{[a(x,u_n,\nabla u_n) - a(x,u,\nabla T_k(u))]\cdot\nabla(u_n-T_k(u))\}^\theta .
\end{aligned}$$

Let us now estimate the first term of the right-hand side of the previous inequality using Hölder's inequality with exponent $1/\theta$: in this way

$$\begin{aligned}
&\int_{\Omega} \{[a(x,u_n,\nabla u_n) - a(x,u,\nabla T_k(u))]\cdot\nabla T_j(u_n-T_k(u))\}^\theta \\
&\le \left(\int_{\Omega} \{[a(x,u_n,\nabla u_n) - a(x,u,\nabla T_k(u))]\cdot\nabla T_j(u_n-T_k(u))\}\right)^\theta \mathrm{meas}(\Omega)^{1-\theta} .
\end{aligned}$$

On the other hand, we can estimate the second term of the right-hand side using the same argument used for

$$\int_{A_k} |[a(x,u_n,\nabla u_n) - a(x,u,\nabla u)]\cdot\nabla(u_n-u)|^\theta :$$

we get

$$\int_{V_j'} |[a(x,u_n,\nabla u_n) - a(x,u,\nabla T_k(u))] \cdot \nabla(u_n - T_k(u))|^\theta \le C\,\mathrm{meas}(V_j')^{\frac{1}{2}}.$$

Summarizing we have

$$\begin{aligned}
&\int_{C_k} \{[a(x,u_n,\nabla u_n) - a(x,u,\nabla u)] \cdot \nabla(u_n - u)\}^\theta \\
&\le C\,\mathrm{meas}(V_j')^{\frac{1}{2}} \\
&\quad + \left(\int_\Omega \{[a(x,u_n,\nabla u_n) - a(x,u,\nabla T_k(u))] \cdot \nabla T_j(u_n - T_k(u))\}\right)^\theta \mathrm{meas}(\Omega)^{1-\theta}.
\end{aligned} \tag{11.2.8}$$

The use of $T_j(u_n - T_k(u))$ as a test function in (11.2.1) yields

$$\int_\Omega a(x,u_n,\nabla u_n) \cdot \nabla T_j(u_n - T_k(u)) = \int_\Omega f_n T_j(u_n - T_k(u)).$$

Estimate (11.2.8) thus implies

$$\begin{aligned}
&\int_{C_k} \{[a(x,u_n,\nabla u_n) - a(x,u,\nabla u)] \cdot \nabla(u_n - u)\}^\theta \\
&\le C\left(\int_\Omega \{f_n T_j(u_n - T_k(u)) - a(x,u,\nabla T_k(u)) \cdot \nabla T_j(u_n - T_k(u))\}\right)^\theta \\
&\quad + C\,\mathrm{meas}(V_j')^{\frac{1}{2}}.
\end{aligned} \tag{11.2.9}$$

Lebesgue's theorem gives

$$\lim_{n\to\infty} \int_\Omega f_n T_j(u_n - T_k(u)) = \int_\Omega f T_j(u - T_k(u)).$$

On the other hand, since u_n converges to u in measure, one has

$$\lim_{n\to\infty} \mathrm{meas}(V_j') = \mathrm{meas}(\{|u - T_k(u)| \ge j\}).$$

We claim that, as $n \to \infty$,

$$\int_\Omega a(x,u,\nabla T_k(u)) \cdot \nabla T_j(u_n - T_k(u)) \to \int_\Omega a(x,u,\nabla T_k(u)) \cdot \nabla T_j(u - T_k(u)).$$

It suffices to prove that $T_j(u_n - T_k(u)) \to T_j(u - T_k(u))$ weakly in $H_0^1(\Omega)$. To this end we observe that

$$\int_{\{|u_n - T_k(u)|<j\}} |\nabla u_n - \nabla T_k(u)|^2 \le 2 \int_{\{|u_n|<j+k\}} |\nabla u_n|^2 + 2 \int_{\{|u_n - T_k(u)|<j\}} |\nabla T_k(u)|^2 . \tag{11.2.10}$$

Now, considering $T_{j+k}(u_n)$ as a test function in (11.2.1), one has

$$\alpha \int_\Omega |\nabla T_{j+k}(u_n)|^2 \le \int_\Omega a(x, u_n, \nabla u_n) \cdot \nabla T_{j+k}(u_n) = \int_\Omega f_n T_{j+k}(u_n) \le \int_\Omega |f|(j+k) ; \tag{11.2.11}$$

consequently

$$\int_{\{|u_n|<j+k\}} |\nabla u_n|^2 \le \int_\Omega |f|(j+k) .$$

By (11.2.10), this yields that the sequence $\|\nabla T_j(u_n + T_k(u))\|^2_{L^2(\Omega)}$ is bounded; since $T_j(u_n - T_k(u)) \to T_j(u - T_k(u))$ a.e. in Ω, $T_j(u_n - T_k(u)) \to T_j(u - T_k(u))$ weakly in $H_0^1(\Omega)$. We can therefore write, by using (11.2.9)

$$\begin{aligned}
&\lim_{n\to\infty} \int_{C_k} |[a(x, u_n, \nabla u_n) - a(x, u, \nabla u)] \cdot \nabla(u_n - u)|^\theta \\
&\quad \le C\,[\text{meas}(\{|u - T_k(u)| \ge j\})]^{\frac12} \\
&\qquad + C \left(\int_\Omega |[f\, T_j(u - T_k(u)) - a(x, u, \nabla T_k(u)) \cdot \nabla T_j(u - T_k(u))| \right)^\theta .
\end{aligned}$$

Step III: We have

$$\begin{aligned}
&\lim_{n\to\infty} \int_\Omega |[a(x, u_n, \nabla u_n) - a(x, u, \nabla u)] \cdot \nabla(u_n - u)|^\theta \\
&\quad \le \lim_{n\to\infty} \int_{A_k} |[a(x, u_n, \nabla u_n) - a(x, u, \nabla u)] \cdot \nabla(u_n - u)|^\theta \\
&\qquad + \lim_{n\to\infty} \int_{C_k} |[a(x, u_n, \nabla u_n) - a(x, u, \nabla u)] \cdot \nabla(u_n - u)|^\theta \\
&\quad \le C\,[\text{meas}(\{|u - T_k(u)| \ge j\})]^{\frac12} + C\,[\text{meas}(\{|u| > k\})]^{\frac12} \\
&\qquad + C \left(\int_\Omega |[f\, T_j(u - T_k(u)) - a(x, u, \nabla T_k(u)) \cdot \nabla T_j(u - T_k(u))| \right)^\theta .
\end{aligned}$$

The last three terms go to 0, as $k \to +\infty$ and so the previous inequality proves (11.2.7).

Lemma 5.8 implies that $\nabla u_n(x) \to \nabla u(x)$ a.e. in Ω. □

11.3 Distributional solutions

In this section, we study the existence of distributional solutions to problem (11.1.1).

Theorem 11.4. *Let $f \in L^m(\Omega)$, with $1 < m < \frac{2N}{N+2}$. Then there exists a distributional solution $u \in W_0^{1,m^*}(\Omega)$ to problem* (11.1.1). *If $f \in L^1(\Omega)$, $u \in W_0^{1,q}(\Omega)$, for every $q < \frac{N}{N-1}$.*

Proof. We distinguish the cases $m > 1$ and $m = 1$. We pass to the limit in problems (11.2.1). In the whole proof C denotes a constant independent on n.

Step I: Let $m > 1$. Let u be the solution found in Lemma 11.3. Hölder's inequality with exponent $\frac{m^*}{r}$, for $r < m^*$, and Lemma 11.2 imply that

$$\int_E |\nabla u_n|^r \le \left(\int_E |\nabla u_n|^{m^*}\right)^{\frac{r}{m^*}} \operatorname{meas}(E)^{1-\frac{r}{m^*}} \le C \operatorname{meas}(E)^{1-\frac{r}{m^*}},$$

where E is any measurable subset of Ω. By Vitali's theorem (Theorem 3.2) $\nabla u_n \to \nabla u$ in $L^r(\Omega)$ for every $r < m^*$. Moreover, from the fact that $\nabla u_n(x) \to \nabla u(x)$ a.e. in Ω, it follows that

$$a(x, u_n(x), \nabla u_n(x)) \to a(x, u(x), \nabla u(x)) \quad \text{a.e. in} \quad \Omega.$$

The growth assumption on a and Lemma 11.2 imply that the sequence $\|a(x, u_n, \nabla u_n)\|_{L^{m^*}(\Omega)}$ is bounded. By Theorem 3.1 $a(x, u_n, \nabla u_n) \to a(x, u, \nabla u)$ weakly in $(L^r(\Omega))^N$. It is now sufficient to pass to the limit in (11.2.1) to prove the result.

Step II: Let $m = 1$. Let u be the function found in Lemma 11.3. Using the same arguments as in *Step I*, $\nabla u_n \to \nabla u$ in $L^q(\Omega)$ for every $q < \frac{N}{N-1}$. This implies that

$$a(x, u_n(x), \nabla u_n(x)) \to a(x, u(x), \nabla u(x)) \quad \text{a.e. in} \quad \Omega.$$

The growth assumption on a and Lemma 11.1 give that the sequence $\|a(x, u_n, \nabla u_n)\|_{L^q(\Omega)}$ is bounded. By Theorem 3.1 $a(x, u_n, \nabla u_n) \to a(x, u, \nabla u)$ weakly in $(L^q(\Omega))^N$. It is now sufficient to pass to the limit in (11.2.1). □

11.4 The linear case: a different proof

In this section, we present a different proof of Theorem 11.4 for linear problems. We prove the following result (see [49]):

Proposition 11.5. *Let M be a $N \times N$ symmetric matrix with bounded coefficients; assume that there exists $\alpha > 0$ such that*

$$M(x)\xi \cdot \xi \ge \alpha |\xi|^2, \quad \forall \xi \in \mathbb{R}^N.$$

Let $f \in L^1(\Omega)$. Then there exists a distributional solution $u \in W_0^{1,q}(\Omega)$ to problem

$$\begin{cases} -\mathrm{div}(M(x)\nabla u) = f\,, & \text{in } \Omega\,, \\ u = 0\,, & \text{on } \partial\Omega\,, \end{cases} \tag{11.4.1}$$

satisfying for every $q < \frac{N}{N-1}$ the following estimate:

$$||u||_{W_0^{1,q}(\Omega)} \le C||f||_{L^1(\Omega)}\,, \tag{11.4.2}$$

for some $C = C(N, q, \alpha) > 0$.

Proof. Let $f_n = T_n(f)$ and $u_n \in H_0^1(\Omega)$ be the weak solution to problem

$$\begin{cases} -\mathrm{div}(M(x)\nabla u_n) = f_n\,, & \text{in } \Omega\,, \\ u_n = 0\,, & \text{on } \partial\Omega\,. \end{cases} \tag{11.4.3}$$

Theorem 6.6 guarantees that $u_n \in L^\infty(\Omega)$, for every $n \in \mathbb{N}$.

Define the following "dual" problem: for every n, let w_n be the solution to

$$\int_\Omega M(x)\nabla w_n \cdot \nabla v = \int_\Omega \chi_k\,|\nabla u_n|^{q-2}\,\nabla u_n \cdot \nabla v\,, \tag{11.4.4}$$

for every $v \in H_0^1(\Omega)$, where χ_k stands for the characteristic function of $\{|\nabla u_n| \le k\}$. We observe that $\chi_k|\nabla u_n|^{q-2}\nabla u_n \in L^\infty(\Omega)$, for every n. Therefore problem (11.4.4) has a solution by Leray–Lions theorem (Theorem 5.1). Moreover we can use Theorem 6.14 to say that $w_n \in L^\infty(\Omega)$. Let us now consider $\varphi = w_n$ as a test function in problem (11.4.3) and $v = u_n$ in ((11.4.4)). In this way

$$\int_\Omega M(x)\nabla u_n \cdot \nabla w_n = \int_\Omega f_n w_n$$

$$\int_\Omega M(x)\nabla w_n \cdot \nabla u_n = \int_\Omega \chi_k\,|\nabla u_n|^{q-2}\,\nabla u_n \cdot \nabla u_n\,.$$

The symmetry of M gives

$$\int_\Omega \chi_k\,|\nabla u_n|^q = \int_\Omega f_n w_n \le ||f_n||_{L^1(\Omega)}||w_n||_{L^\infty(\Omega)}\,.$$

By Theorem 6.14 applied with $m = \frac{q}{q-1} > N$, we get

$$\int_\Omega \chi_k\,|\nabla u_n|^q \le C||f_n||_{L^1(\Omega)}\left(\int_\Omega \chi_k\,|\nabla u_n|^q\right)^{\frac{q-1}{q}},$$

where C depends on N, q and α. Therefore

$$\left(\int_{\Omega} \chi_k |\nabla u_n|^q\right)^{1/q} \leq C||f_n||_{L^1(\Omega)} . \tag{11.4.5}$$

Passing to the limit as $k \to +\infty$ in (11.4.5), Fatou's lemma implies

$$||\nabla u_n||_{L^q(\Omega)} \leq C||f||_{L^1(\Omega)} ,$$

for every $q \in [1, \frac{N}{N-1})$. Therefore, u_n is bounded in $W_0^{1,q}(\Omega)$ and there exists a subsequence weakly converging to some u in $W_0^{1,q}(\Omega)$. It easily seen that u is a weak solution to problem (11.4.1). Moreover estimate (11.4.2) follows from the weak lower semicontinuity of the L^q norm. □

11.5 Entropy solutions

In this section, we introduce a new notion of solution to problem (11.1.1): the entropy solution, which is a particular distributional solution. We give the definition, some properties and then we prove the existence. We will understand the importance of entropy solutions in Chapter 12 where we will study the uniqueness of solutions.

Definition 11.6. We set

$$\mathcal{T}(\Omega) = \{u : \Omega \to \mathbb{R} \quad \text{measurable and finite a.e. such that} \\ T_k(u) \in H_0^1(\Omega) , \quad \forall\, k > 0\} .$$

Let us study the gradient of a $\mathcal{T}(\Omega)$ function.

Lemma 11.7. *For every $u \in \mathcal{T}(\Omega)$ there exists a unique measurable map $v : \Omega \to \mathbb{R}^N$ such that*

$$\nabla T_k(u) = v\, \chi_{\{|u|<k\}} \quad \text{a.e. in} \quad \Omega .$$

Moreover, if $u \in H_0^1(\Omega)$, v coincides with the usual distributional gradient ∇u.

Proof. Let $k, \varepsilon > 0$; then

$$T_k(T_{k+\varepsilon}(u)) = T_k(u);$$

since $T_j(u)$ belongs $H_0^1(\Omega)$ for every j, one has

$$\nabla T_k(T_{k+\varepsilon}(u)) = \nabla T_k(u) .$$

In $\Omega_k = \{|u| < k\}$ the previous equality is equivalent to $\nabla T_{k+\varepsilon}(u) = \nabla T_k(u)$ for every $\varepsilon > 0$. Since $\bigcup_{k>0} \Omega_k = \Omega$, setting

$$\nabla u(x) = \nabla T_k(u(x)) \quad \text{a.e. in} \quad \Omega_k ,$$

the lemma is proved. □

We are now in position to give the definition of entropy solution to problem (11.1.1).

Definition 11.8. Let f be an $L^1(\Omega)$ function. A $\mathcal{T}(\Omega)$ function u is an entropy solution to problem (11.1.1) if

$$\int_\Omega a(x,u,\nabla u)\cdot\nabla T_k(u-\varphi)\le\int_\Omega f\,T_k(u-\varphi)$$
$$\forall k>0\quad\text{and}\quad\forall\varphi\in H_0^1(\Omega)\cap L^\infty(\Omega)\,.\tag{11.5.1}$$

Remark 11.9. The entropy inequality (11.5.1) is well defined. Indeed, the right-hand side is finite, since $T_k(u-\varphi)$ is bounded. As for the left one, we observe that $T_k(u-\varphi)$ belongs to $H_0^1(\Omega)$ and

$$\nabla T_k(u-\varphi)=\nabla(u-\varphi)\,\chi_{\{|u-\varphi|\le k\}}\,.$$

Therefore

$$\int_\Omega a(x,u,\nabla u)\cdot\nabla T_k(u-\varphi)=\int_{\{|u-\varphi|\le k\}} a(x,u,\nabla u)\cdot\nabla(u-\varphi)\,.$$

In the set $\{|u-\varphi|\le k\}$, ∇u belongs to $L^2(\Omega)$ and consequently $a(x,u,\nabla u)$ belongs to $L^2(\Omega)$ too. The left-hand side is thus well defined.

Remark 11.10. A $H_0^1(\Omega)$ function u satisfying

$$\int_\Omega a(x,u,\nabla u)\cdot\nabla v=\int_\Omega f\,v\,,\qquad\forall\,v\in H_0^1(\Omega)$$

is clearly an entropy solution to problem (11.1.1). This implies that the $H_0^1(\Omega)$ distributional solutions found in Chapter 5 (when the source $f\in L^m(\Omega)$, $m\ge\frac{2N}{N+2}$), are entropy solutions to problem (11.1.1).

Let us study the main properties of the entropy solutions.

Proposition 11.11. *Let $u\in\mathcal{T}(\Omega)$ be an entropy solution to problem* (11.1.1). *Then u belongs to $M^{\frac{N}{N-2}}(\Omega)$.*

Proof. In the whole proof C will denote a constant independent of u and k. Consider $\varphi=0$ in (11.5.1): the ellipticity of a implies

$$\alpha\|\nabla T_k(u)\|^2_{L^2(\Omega)}\le\int_\Omega a(x,u,\nabla u)\cdot\nabla T_k(u)\le k\,\|f\|_{L^1(\Omega)}\,,\qquad\forall k>0\,.$$

Let $0<h<k$. By Sobolev's inequality on the left-hand side, the above estimate gives

$$h^{2^*}\,\text{meas}(\{|u|>h\})<\int_{\{|u|\le k\}}|u|^{2^*}\le C\,k^{2^*/2}\,.$$

Now, let $h = \frac{k}{2}$. We deduce that

$$\text{meas}(\{|u| > h\}) \le C h^{-\frac{N}{N-2}},$$

meaning, $u \in M^{\frac{N}{N-2}}(\Omega)$. □

Remark 11.12. The same technique can be used to prove the following statement. Assume that u_n is a sequence of functions such that for a positive constant C it holds

$$\|\nabla T_k(u_n)\|^2_{L^2(\Omega)} \le C k, \qquad \forall k > 0.$$

Then there exists a positive constant c (independent on n and k) such that

$$\text{meas}(\{|u_n| > k\}) \le \frac{c}{k^{\frac{N}{N-2}}}.$$

Proposition 11.13. *Let u be an entropy solution to problem* (11.1.1). *Then* $|\nabla u| \in M^{\frac{N}{N-1}}(\Omega)$.

Proof. In the proof C will denote a constant independent of u and k. As in the previous proposition, consider $\varphi = 0$ in (11.5.1). We get

$$\int_\Omega |\nabla T_k(u)|^2 \le C k.$$

Therefore

$$t^2 \,\text{meas}(\{|u| \le k, |\nabla u| > t\}) \le \int_{\{|u|\le k, |\nabla u|>t\}} |\nabla u|^2 \le C k.$$

The last estimate and Proposition 11.11 give

$$\text{meas}(\{|u| \le k, |\nabla u| > t\}) + \text{meas}(\{|u| > k\}) \le C \frac{k}{t^2} + \frac{C}{k^{\frac{N}{N-2}}}.$$

Consequently

$$\text{meas}(\{|\nabla u| > t\}) \le C \frac{k}{t^2} + \frac{C}{k^{\frac{N}{N-2}}};$$

minimizing with respect to k the function

$$k \to \frac{k}{t^2} + \frac{1}{k^{\frac{N}{N-2}}},$$

one has

$$\text{meas}(\{|\nabla u| > t\}) \le \frac{C}{t^{\frac{N}{N-1}}}.$$ □

Remark 11.14. With the same technique one can prove the following result. Assume that u_n is a sequence of functions such that for a positive constant C it holds

$$\|\nabla T_k(u_n)\|^2_{L^2(\Omega)} \le C\,k\,, \qquad \forall\, k>0\,.$$

Then there exists a positive constant c such that

$$\text{meas}(\{|\nabla u_n|>k\}) \le \frac{c}{k^{\frac{N}{N-1}}}\,.$$

Corollary 11.15. *Let $u \in \mathcal{T}(\Omega)$ be an entropy solution to problem* (11.1.1). *Then $|a(x,u,\nabla u)| \in L^1(\Omega)$.*

Proof. The growth assumption on a implies

$$\text{meas}(\{|a(x,u,\nabla u)|>k\}) \le \text{meas}(\{|u|+|\nabla u|>k\})\,. \tag{11.5.2}$$

The above results give $|u|+|\nabla u| \in M^{\frac{N}{N-1}}(\Omega)$. We deduce from (11.5.2) that

$$|a(x,u,\nabla u)| \in M^{\frac{N}{N-1}}(\Omega) \subseteq L^1(\Omega)\,.$$

□

We now prove that the entropy solutions are distributional solutions.

Proposition 11.16. *Let $u \in \mathcal{T}(\Omega)$ be an entropy solution to problem* (11.1.1). *Then u is a distributional solution, that is,*

$$\int_\Omega a(x,u,\nabla u)\cdot\nabla\psi = \int_\Omega f\,\psi\,, \qquad \forall\,\psi \in C_0^\infty(\Omega)\,.$$

Proof. Let $\psi \in C_0^\infty(\Omega)$; consider $\varphi = T_h(u) - \psi \in H_0^1(\Omega)\cap L^\infty(\Omega)$ in (11.5.1). Then

$$\int_{\{|u-T_h(u)+\psi|<k\}} a(x,u,\nabla u)\cdot(\nabla u\,\chi_{\{|u|>h\}} + \nabla\psi) \le \int_\Omega f\,T_k(u-T_h(u)+\psi)\,.$$

The ellipticity of a gives

$$\int_{\{|u-T_h(u)+\psi|<k\}} a(x,u,\nabla u)\cdot\nabla\psi \le \int_\Omega f\,T_k(u-T_h(u)+\psi)\,.$$

If we consider $k > \|\psi\|_{L^\infty(\Omega)}$ we have

$$\chi_{\{|u-T_h(u)+\psi|<k\}} \to \chi_\Omega\,, \quad \text{as } h\to+\infty\,.$$

Since $a(x,u,\nabla u)\cdot\nabla\psi \in L^1(\Omega)$ we can use Lebesgue's theorem and prove, passing to the limit as $h\to+\infty$, that

$$\int_\Omega a(x,u,\nabla u)\cdot\nabla\psi \le \int_\Omega f\,\psi\,, \quad \forall\,\psi\in C_0^\infty(\Omega)\,.$$

The choice of $-\psi$ instead of ψ gives the opposite inequality. Therefore u is a distributional solution. □

We can now prove the existence of entropy solutions to problem (11.1.1):

Theorem 11.17. *Let $f \in L^m(\Omega)$, $m \geq 1$. Then there exists an entropy solution $u \in \mathcal{T}(\Omega)$ to problem* (11.1.1).

(1) *If $f \in L^1(\Omega)$, u belongs to $M^{\frac{N}{N-2}}(\Omega)$, $|\nabla u| \in M^{\frac{N}{N-1}}(\Omega)$.*

(2) *If $f \in L^m(\Omega)$, $m > 1$ then u belongs to $W_0^{1,m^*}(\Omega)$.*

Proof. Let u_n be the solutions to problems (11.2.1). Let us consider $T_k(u_n - \varphi)$ as a test function. By Lemma 11.1, u_n converges weakly to some u in $W_0^{1,q}(\Omega)$ with $q < \frac{N}{N-1}$, up to a subsequence. By Lemma 11.3 $\nabla u_n \to \nabla u$ a.e. in Ω and so $a(x, u_n, \nabla u_n) \to a(x, u, \nabla u)$ a.e. in Ω. On the other hand, for every k and $\varphi \in H_0^1(\Omega) \cap L^\infty(\Omega)$, $\nabla T_k(u_n - \varphi) \to \nabla T_k(u - \varphi)$ a.e. in Ω. Fatou's Lemma implies that

$$\liminf_{n\to\infty} \int_\Omega a(x, u_n, \nabla u_n) \cdot \nabla T_k(u_n - \varphi) \geq \int_\Omega a(x, u, \nabla u) \cdot \nabla T_k(u - \varphi).$$

Lebesgue's theorem yields

$$\lim_{n\to\infty} \int_\Omega f_n\, T_k(u_n - \varphi) = \int_\Omega f\, T_k(u - \varphi).$$

We have thus proved that u satisfies (11.5.1). To prove that $T_k(u)$ belongs to $H_0^1(\Omega)$ for every $k > 0$, one can choose $T_k(u_n)$ in (11.2.1), as a test function. By the ellipticity of a, it is easy to see that

$$\alpha \|\nabla T_k(u_n)\|^2_{L^2(\Omega)} \leq k \|f\|_{L^1(\Omega)}.$$

Since $\nabla u_n \to \nabla u$ a.e. in Ω, passing to the limit as $n \to \infty$ we have that

$$\alpha \|\nabla T_k(u)\|^2_{L^2(\Omega)} \leq k \|f\|_{L^1(\Omega)}.$$

Thus $T_k(u)$ belongs to $H_0^1(\Omega)$ for every $k > 0$.

As for the summability of u:

(1) if f belongs to $L^1(\Omega)$, it is sufficient to use Propositions 11.11 and 11.13;

(2) if f belongs to $L^1(\Omega)$, $m > 1$, one can use Lemma 11.2. □

Remark 11.18. We observe that similar results can be proved (see [12, 13]) in the case where a satisfies the coercivity assumption

$$a(x, s, \xi) \cdot \xi \geq \alpha |\xi|^p, \quad \forall\, \xi \in \mathbb{R}^N.$$

Indeed, if the source $f \in L^1(\Omega)$, $1 < p < N$, then there exists an entropy solution $u \in M^{p_1}(\Omega)$ with $|\nabla u| \in M^{p_2}(\Omega)$, where $p_1 = \frac{N(p-1)}{N-p}$, $p_2 = \frac{N(p-1)}{N-1}$. If $p > 2 - \frac{1}{N}$ then there exists a distributional solution $u \in W_0^{1,q}(\Omega)$, $q < p_2$.

11.6 A comparison between entropy solutions and distributional solutions

In the previous sections we have proved the existence of entropy solutions and distributional solutions to problem (11.1.1). We now want to compare these two notions, showing that they are not equivalent.

Proposition 11.16 establishes that an entropy solution to problem (11.1.1) is a distributional solution. We now present a linear problem having a distributional solution which is not an entropy solution (according to an idea given originally by Serrin in [47] and presented differently in [44]).

Let $M = (m_{ij})_{i,j}^{N} \in \mathbb{R}^{N\times N}$ be the following matrix: if $x = (x_1, .., x_N)$

$$\begin{cases} m_{ij} = \delta_{ij} + \frac{1-\varepsilon^2}{\varepsilon^2}\frac{x_i x_j}{x_1^2+x_2^2}, & i,j = 1,2 \\ m_{ij} = \delta_{ij}, & i,j \neq 1,2. \end{cases} \tag{11.6.1}$$

M is elliptic and bounded. We claim that

$$u_0(x) = \frac{x_1}{(x_1^2+x_2^2)^{\frac{1+\varepsilon}{2}}}, \qquad \varepsilon < \frac{1}{N-1} \tag{11.6.2}$$

satisfies $-\operatorname{div}(M(x)\nabla u_0) = 0$ in $\mathbb{R}^N$. Let φ be a C^∞ function with compact support in $\mathbb{R}^N$. Integrating by parts yields

$$\begin{aligned} \int_{\mathbb{R}^N} M(x)\nabla u_0 \cdot \nabla\varphi &= \lim_{\rho\to 0} \int_{\{|x|\geq\rho\}} M(x)\nabla u_0 \cdot \nabla\varphi \\ &= -\lim_{\rho\to 0} \oint_{\{|x|=\rho\}} \varphi M(x)\nabla u_0 \cdot \frac{x}{|x|} ds \\ &= -\frac{1}{\varepsilon}\lim_{\rho\to 0} \oint_{\{|x|=\rho\}} \frac{1}{\rho^\varepsilon}\varphi x_1 ds \\ &= 0 \end{aligned}$$

since the last integral is $o(\rho^2)$ as $\rho \to 0$. Moreover, setting $\Omega = \{|x| < 1\}$, one has that $u_0 \in W^{1,q}(\Omega)$, $q < \frac{N}{N-1}$ and $T_k(u_0) \notin H^1(\Omega)$. Consequently, if

$$f(x) = \frac{1-\varepsilon^2}{\varepsilon^2}\frac{x_1}{x_1^2+x_2^2}, \tag{11.6.3}$$

the function

$$v(x) = x_1 - u_0(x) \tag{11.6.4}$$

is a distributional solution to problem

$$\begin{cases} -\operatorname{div}(M(x)\nabla v) = f, & \text{in } \Omega, \\ v = 0, & \text{on } \partial\Omega, \end{cases}$$

but it is not an entropy solution, since it does not belong to $\mathcal{T}(\Omega)$.

11.7 Measure sources

In this section, we study the existence of distributional solutions to

$$\begin{cases} -\operatorname{div}(a(x,u,\nabla u)) = \mu\,, & \text{in } \Omega\,, \\ u = 0\,, & \text{on } \partial\Omega\,, \end{cases} \tag{11.7.1}$$

where μ is a measure, that is, $\mu \in \mathcal{M}(\Omega)$ (see the appendix for the definition and some properties). We will prove the existence of a $W_0^{1,1}(\Omega)$ function such that

$$\int_\Omega a(x,u,\nabla u)\cdot\nabla\varphi = \int_\Omega \varphi d\mu\,, \quad \forall\,\varphi \in C_0^\infty(\Omega)\,.$$

We will use the approximating problems (11.2.1) where f_n is a sequence of $C_0^\infty(\Omega)$ functions such that $\|f_n\|_{L^1(\Omega)} \le \|\mu\|_{\mathcal{M}(\Omega)}$ and $f_n \to \mu$ $*$-weakly in $\mathcal{M}(\Omega)$ (Theorem 11.39 guarantees the existence of such a sequence).

Theorem 11.19. *Let $\mu \in \mathcal{M}(\Omega)$. Then there exists a distributional solution $u \in W_0^{1,q}(\Omega)$, $q < \frac{N}{N-1}$, to problem* (11.7.1).

As in the case where the source is in $L^1(\Omega)$, we need the following two lemmata:

Lemma 11.20. *Let $\mu \in \mathcal{M}(\Omega)$. Then the sequence of solutions u_n to* (11.2.1) *is bounded in $W_0^{1,q}(\Omega)$, $q < \frac{N}{N-1}$.*

Proof. The proof is similar to that one of Lemma 11.1. Indeed, considering $v_n = [(1+|u_n|)^{2\lambda-1} - 1]\operatorname{sgn}(u_n)$, $\lambda = \frac{q(N-2)}{2(N-q)}$ as test functions, one has the following estimate for the right-hand side

$$\int_\Omega f_n v_n \le \int_\Omega |f_n| \le \|\mu\|_{\mathcal{M}(\Omega)}\,,$$

in the same spirit as (11.2.2). We can then use the same arguments as in Lemma 11.1 to prove that the solutions u_n to problems (11.2.1) are bounded in $W_0^{1,q}(\Omega)$, $q < \frac{N}{N-1}$. □

Lemma 11.21. *Let $\mu \in \mathcal{M}(\Omega)$. Then, if u_n are the $H_0^1(\Omega)$ solutions to problems* (11.2.1), *there exists a function $u \in W_0^{1,q}(\Omega)$, $q < \frac{N}{N-1}$ such that $\nabla u_n \to \nabla u$ a.e. in Ω, up to a subsequence.*

Proof. The proof is not very different from that one of Lemma 11.3. C will denote a constant independent on n. Step I follows in the same way:

$$\int_{A_k} |[a(x,u_n,\nabla u_n) - a(x,u,\nabla u)]\cdot\nabla(u_n - u)|^\theta \le C\operatorname{meas}(A_k)^{1/2}\,,$$

where $A_k = \{|u_n| \ge k\}$. As for Step *II* one can follow the same arguments, except for (11.2.9):

$$\int_{C_k} |[a(x,u_n,\nabla u_n) - a(x,u,\nabla u)] \cdot \nabla(u_n - u)|^\theta$$

$$\le C \left(\int_\Omega \{f_n T_j(u_n - T_k(u)) - a(x,u,\nabla T_k(u)) \cdot \nabla T_j(u_n - T_k(u))\} \right)^\theta$$

$$+ C \operatorname{meas}(V_j')^{\frac{1}{2}},$$

with $V_j' = \{|u_n - T_k(u)| > j\}$ and $C_k = \{|u_n| \le k\}$. Regarding the first term of the right-hand side we observe that

$$\left(\int_\Omega \{f_n T_j(u_n - T_k(u)) - a(x,u,\nabla T_k(u)) \cdot \nabla T_j(u_n - T_k(u))\} \right)^\theta$$

$$\le C \left| \int_\Omega f_n T_j(u_n - T_k(u)) \right|^\theta + C \left| \int_\Omega a(x,u,\nabla T_k(u)) \cdot \nabla T_j(u_n - T_k(u)) \right|^\theta$$

$$\le C \left(\int_\Omega j|f_n| \right)^\theta + C \left| \int_\Omega a(x,u,\nabla T_k(u)) \cdot \nabla T_j(u_n - T_k(u)) \right|^\theta$$

$$\le C j^\theta \|\mu\|^\theta_{\mathcal{M}(\Omega)} + C \left| \int_\Omega a(x,u,\nabla T_k(u)) \cdot \nabla T_j(u_n - T_k(u)) \right|^\theta .$$

Therefore

$$\int_{C_k} |[a(x,u_n,\nabla u_n) - a(x,u,\nabla u)] \cdot \nabla(u_n - u)|^\theta$$

$$\le C j^\theta \|\mu\|^\theta_{\mathcal{M}(\Omega)} + C \left| \int_\Omega a(x,u,\nabla T_k(u)) \cdot \nabla T_j(u_n - T_k(u)) \right|^\theta + C \operatorname{meas}(V_j')^{\frac{1}{2}}.$$

One can then use the same arguments as in Lemma 11.3 for the second and the third term of the last estimate. In conclusion

$$\lim_{n\to\infty} \int_\Omega \{[a(x,u_n,\nabla u_n) - a(x,u,\nabla u)] \cdot \nabla(u_n - u)\}^\theta$$

$$\le C j^\theta + C\,[\operatorname{meas}(\{|u - T_k(u)| \ge j\})]^{\frac{1}{2}} + C\,[\operatorname{meas}(\{|u| > k\})]^{\frac{1}{2}}$$

$$+ C \left(\int_\Omega |[f\, T_j(u - T_k(u)) - a(x,u,\nabla T_k(u)) \cdot \nabla T_j(u - T_k(u))| \right)^\theta .$$

As $k \to \infty$, the last three terms tend to 0. As $j \to 0$, one has

$$\lim_{n\to\infty} \int_\Omega \{[a(x, u_n, \nabla u_n) - a(x, u, \nabla u)] \cdot \nabla(u_n - u)\}^\theta = 0;$$

this implies that, up to a subsequence, $\nabla u_n \to \nabla u$ a.e. in Ω. □

Using the previous lemmata one can prove Theorem 11.19 in the same way as Theorem 11.4.

11.8 The regularizing effects of a lower order term

In this section, we are going to study the following problem for $p \geq 1$:

$$\begin{cases} -\mathrm{div}(a(x, u, \nabla u)) + |u|^{p-1}u = f, & \text{in } \Omega, \\ u = 0, & \text{on } \partial\Omega, \end{cases} \tag{11.8.1}$$

proving that the lower order term has some regularizing effect on the solutions and on their gradient. This phenomenon was already observed in [23] and [24] for the solutions of

$$\begin{cases} -\Delta u + |u|^{p-1}u = f, & \text{in } \Omega, \\ u = 0, & \text{on } \partial\Omega : \end{cases}$$

indeed if u is a solution, then $|u|^p$ belongs to $L^m(\Omega)$ under the assumption $f \in L^m(\Omega)$, $m \geq 1$.

We will prove the following result (see [16]) on the existence of distributional solutions to problem (11.8.1).

Theorem 11.22. *Let $f \in L^m(\Omega)$, $m \geq 1$.*
(1) *If $m > 1$ and $p \geq \frac{1}{m-1}$, there exists a distributional solution $u \in H_0^1(\Omega)$.*
(2) *If $m > 1$ and $p < \frac{1}{m-1}$, there exists a distributional solution $u \in W_0^{1,q}(\Omega)$, for every $q < \frac{2pm}{1+pm}$.*
(3) *If $m = 1$, there exists a distributional solution $u \in W_0^{1,q}(\Omega)$, for every $q < \frac{2p}{1+p}$.*

As for problem (11.1.1) we will work by approximation, considering the following family of approximating problems:

$$\begin{cases} -\mathrm{div}(a(x, u_n, \nabla u_n)) + |T_n(u_n)|^{p-1}T_n(u_n) = T_n(f), & \text{in } \Omega, \\ u_n = 0, & \text{on } \partial\Omega. \end{cases} \tag{11.8.2}$$

The existence of solutions $u_n \in H_0^1(\Omega)$, for every n, follows from Theorem 5.1, since $|T_n(s)|^p \leq n^p$. In the following lemmata we prove some uniform estimates on the solutions u_n to problems (11.8.2).

Lemma 11.23. *Let $u_n \in H_0^1(\Omega)$ be the solutions to problems* (11.8.2). *The following estimate holds for every $n \in \mathbb{N}$:*

$$\left[\int_\Omega |T_n(u_n)|^{pm}\right]^{\frac{1}{m}} \le \|f\|_{L^m(\Omega)}\,. \tag{11.8.3}$$

Proof. In the case $m > 1$ we choose $|T_n(u_n)|^{p(m-1)}\operatorname{sgn}(u_n)$ as a test function in (11.8.2). Dropping the operator term of the left-hand side, and using Hölder's inequality on the right one, we obtain

$$\int_\Omega |T_n(u_n)|^{pm} \le \int_\Omega |f||T_n(u_n)|^{p(m-1)} \le \|f\|_{L^m(\Omega)}\left[\int_\Omega |T_n(u_n)|^{pm}\right]^{1-\frac{1}{m}}.$$

This implies (11.8.3). In the case $m = 1$ we choose $\frac{T_k(u_n)}{k}$ as a test function in (11.8.2). Again, dropping the operator term from the left-hand side, one has

$$\int_\Omega |T_n(u_n)|^{p-1}T_n(u_n)\frac{T_k(u_n)}{k} \le \int_\Omega |f|\,.$$

It is now sufficient to pass to the limit as $k \to 0$ and use Fatou's lemma. □

Lemma 11.24. *Let $m > 1$. Let u_n be the solutions to problems* (11.8.2).
(1) *If $p \ge \frac{1}{m-1}$, then the sequence $T_n(u_n)$ is bounded in $H_0^1(\Omega)$.*
(2) *If $p < \frac{1}{m-1}$, then the sequence $T_n(u_n)$ is bounded in $W_0^{1,q}(\Omega)$, with $q < \frac{2pm}{1+pm}$.*

Proof. If $p \ge \frac{1}{m-1}$ we consider $T_n(u_n)$ as a test function in (11.8.2). Then by the ellipticity of a and Hölder's inequality on the right-hand side we get

$$\alpha\int_\Omega |\nabla T_n(u_n)|^2 \le \int_\Omega T_n(f)T_n(u_n) \le \left[\int_\Omega |T_n(u_n)|^{pm}\right]^{\frac{1}{pm}}\left[\int_\Omega |f|^{(pm)'}\right]^{\frac{1}{(pm)'}}.$$

The last term is finite, due to Lemma 11.23.

In the case $p < \frac{1}{m-1}$ we consider $v_n = [(1+|T_n(u_n)|)^{2\lambda-1}-1]\operatorname{sgn}(u_n)$, $\lambda < \frac{1}{2}$, as a test function in (11.8.2). Then, dropping the lower order term, one can prove the following estimate, using the same arguments as in Lemma 11.1:

$$\int_\Omega \frac{|\nabla T_n(u_n)|^2}{(1+|T_n(u_n)|)^{2(1-\lambda)}} \le \frac{\|f\|_{L^1(\Omega)}}{\alpha(1-2\lambda)}\,. \tag{11.8.4}$$

On the other hand, let $q < 2$. We can write

$$\int_\Omega |\nabla T_n(u_n)|^q = \int_\Omega \frac{|\nabla T_n(u_n)|^q}{(1+|T_n(u_n)|)^{2(1-\lambda)\frac{q}{2}}}(1+|T_n(u_n)|)^{2(1-\lambda)\frac{q}{2}}\,.$$

Using the Hölder inequality with exponent $2/q$ on the right-hand side, we get

$$\int_\Omega |\nabla T_n(u_n)|^q \le \left(\int_\Omega \frac{|\nabla T_n(u_n)|^2}{(1+|T_n(u_n)|)^{2(1-\lambda)}}\right)^{\frac{q}{2}} \left(\int_\Omega (1+|T_n(u_n)|)^{\frac{(1-\lambda)2q}{2-q}}\right)^{1-\frac{q}{2}}.$$

By (11.8.4) the last term is bounded if $\frac{(1-\lambda)2q}{2-q} \le pm$, that is, $\lambda \ge 1 - pm\frac{2-q}{2q}$. This choice of λ is possible: indeed $\frac{1}{2} > 1 - pm\frac{2-q}{2q}$, since $q < \frac{2pm}{1+pm} (< 2)$. □

Lemma 11.25. *Let $f \in L^1(\Omega)$. Let u_n be the solutions to problems* (11.8.2). *Then the sequence $T_n(u_n)$ is bounded in $W_0^{1,q}(\Omega)$, with $q < \frac{2p}{1+p}$.*

Proof. Consider $v_n = [(1+|T_n(u_n)|)^{2\lambda-1} - 1]\operatorname{sgn}(u_n)$, with $\lambda < \frac{1}{2}$, as a test function in (11.8.2). With the same technique as in the previous lemma, one has

$$\int_\Omega |\nabla T_n(u_n)|^q \le \left(\int_\Omega \frac{|\nabla T_n(u_n)|^2}{(1+|T_n(u_n)|)^{2(1-\lambda)}}\right)^{\frac{q}{2}} \left(\int_\Omega (1+|T_n(u_n)|)^{\frac{(1-\lambda)2q}{2-q}}\right)^{1-\frac{q}{2}}.$$

The last term is bounded if $\frac{(1-\lambda)2q}{2-q} \le p$, that is, $\lambda \ge 1 - p\frac{2-q}{2q}$. This choice is possible: indeed $\frac{1}{2} > 1 - p\frac{2-q}{2q}$, since $q < \frac{2p}{1+p} (< 2)$. □

We can now prove Theorem 11.22:

Proof of Theorem 11.22. Assume for instance, that $m > 1$ and $p \ge \frac{1}{m-1}$; the other cases are similar. Let $u_n \in H_0^1(\Omega)$ be the solutions to problems (11.8.2). By Lemma 11.24, there exists a function $u \in H_0^1(\Omega)$ such that $T_n(u_n) \to u$ weakly in $H_0^1(\Omega)$, up to a subsequence. Remark that $u_n \in H_0^1(\Omega)$ satisfies

$$-\operatorname{div}(a(x, u_n, \nabla u_n)) = g_n,$$

where $g_n = f_n - |T_n(u_n)|^{p-1} T_n(u_n)$. Since $g_n \in L^1(\Omega)$ for every fixed n, we can apply Lemma 11.3 to get that $\nabla u_n \to \nabla u$ a.e. in Ω, up to a subsequence. This allows us to pass to the limit in the first term of (11.8.2) as in the proof of Theorem 11.4. The limit of the right-hand side of (11.8.2) is easy.

For the limit of the lower order term it is useful to prove that

$$\int_{\{|T_n(u_n)|>t\}} |T_n(u_n)|^p \le \int_{\{|T_n(u_n)|>t\}} |f|. \tag{11.8.5}$$

Let ψ_i be a sequence of increasing, positive, uniformly bounded $C^\infty(\Omega)$ functions, such that

$$\psi_i(s) \to \begin{cases} 1, & s \ge t, \\ 0, & |s| < t, \\ -1, & s \le -t. \end{cases}$$

Choosing $\psi_i(T_n(u_n))$ in (11.8.2), we get

$$\int_\Omega |T_n(u_n)|^{p-1}T_n(u_n)\,\psi_i(T_n(u_n)) \le \int_\Omega T_n(f)\,\psi_i(T_n(u_n))\,.$$

The limit on i implies (11.8.5). We are now going to prove that if E is any measurable subset of Ω, then

$$\lim_{\text{meas}(E)\to 0}\int_E |T_n(u_n)|^p = 0 \quad \text{uniformly with respect to } n\,.$$

By (11.8.5), for any $t > 0$ we have

$$\int_E |T_n(u_n)|^p \le t^p \,\text{meas}(E) + \int_{E\cap\{|T_n(u_n)|>t\}} |T_n(u_n)|^p \le t^p \,\text{meas}(E) + \int_{\{|T_n(u_n)|>t\}} |f|\,.$$

By Lemma 11.23 the sequence $T_n(u_n)$ is bounded in $M^{mp}(\Omega)$. This and the fact that $f \in L^1(\Omega)$ allow us to say that for any given $\varepsilon > 0$, there exists t_ε such that

$$\int_{\{|T_n(u_n)|>t_\varepsilon\}} |f| \le \varepsilon\,.$$

In this way

$$\int_E |T_n(u_n)|^p \le t_\varepsilon^p \,\text{meas}(E) + \varepsilon$$

and so

$$\lim_{\text{meas}(E)\to 0}\int_E |T_n(u_n)|^p \le \varepsilon\,, \quad \forall \varepsilon > 0\,.$$

We thus proved that $\lim_{\text{meas}(E)\to 0}\int_E |T_n(u_n)|^p = 0$ uniformly with respect to n. Vitali's theorem (Theorem 3.2) implies that $|T_n(u_n)|^{p-1}T_n(u_n) \to |u|^{p-1}u$ in $L^1(\Omega)$. Therefore we can pass to the limit in (11.8.2) to get a solution to problem (11.8.1). □

11.9 T-minima

The last paragraph of this chapter is devoted to the study of integral functionals unbounded from below. More precisely we will be interested to functionals of the form

$$J(v) = \int_\Omega j(x,\nabla v) - \int_\Omega fv$$

where $j : \Omega \times \mathbb{R}^N \to \mathbb{R}$ is a Carathéodory function, convex in the last variable, such that

$$\alpha|\xi|^2 \le j(x,\xi) \le \beta|\xi|^2 \tag{11.9.1}$$

for some positive α and f belongs $L^1(\Omega)$. Due to the summability of f, J is not bounded from below in $H_0^1(\Omega)$. For example, in $\Omega = B(0,1)$, for $\beta < N$, consider

$$J(v) = \int_\Omega |\nabla v|^2 - \int_\Omega \frac{v}{|x|^\beta}.$$

Then the sequence $v_n = T_n(\frac{1}{|x|^\alpha})$, $\alpha > N - \beta$, is such that $J(v_n) \to -\infty$. The nonexistence of minima led to the definition of T-minima, in analogy with the definition of entropy solutions to elliptic problems (see [5] and [33]).

Definition 11.26. Let $f \in L^1(\Omega)$. A function $u \in \mathcal{T}(\Omega)$ is a T-minimum for the functional

$$J(v) = \int_\Omega j(x,\nabla v) - \int_\Omega fv$$

if, for every φ in $H_0^1(\Omega) \cap L^\infty(\Omega)$ and every $k > 0$, we have

$$\int_\Omega j(x,\nabla(\varphi + T_k(u-\varphi)) \le \int_\Omega j(x,\nabla\varphi) + \int_\Omega f\, T_k(u-\varphi). \tag{11.9.2}$$

Observe that if J has a minimum $u \in H_0^1(\Omega)$ (see Chapter 9, in the case where $f \in L^{\frac{2N}{N+2}}(\Omega)$), then u satisfies (11.9.2).

Lemma 11.27. *Let $u_n \in \mathcal{T}(\Omega)$ be a sequence of functions such that for a positive constant C the following estimate holds:*

$$\int_\Omega |\nabla T_k(u_n)|^2 \le Ck.$$

Then there exists $u \in \mathcal{T}(\Omega)$ such that, up to a subsequence, $T_k(u_n) \to T_k(u)$ weakly in $H_0^1(\Omega)$.

Proof. Since the sequence $T_k(u_n)$ is bounded in $H_0^1(\Omega)$, there exists v_k in $L^2(\Omega)$ such that, up to a subsequence $T_k(u_n) \to v_k$ weakly in $H_0^1(\Omega)$ and strongly in $L^2(\Omega)$. By Remark 11.12, there exists a positive constant c such that

$$\operatorname{meas}(\{|u_n| > k\}) \le \frac{c}{k^{\frac{N}{N-2}}}$$

for every $n \in \mathbb{N}$. Now, fix $\varepsilon > 0$. The above estimate allows us to choose $\overline{k}$ such that

$$\operatorname{meas}(\{|u_n| > k\}) \le \frac{\varepsilon}{3}, \quad \operatorname{meas}(\{|u_m| > k\}) \le \frac{\varepsilon}{3} \tag{11.9.3}$$

for every $n, m \in \mathbb{N}$ and for every $k > \overline{k}$. Moreover, since $T_k(u_n)$ is a Cauchy sequence in measure (recall that $T_k(u_n) \to v_k$ in measure, as $n \to \infty$), there exists η_ε such that for every $n, m > \eta_\varepsilon$

$$\operatorname{meas}(\{|T_k(u_n) - T_k(u_m)| > \sigma\}) \leq \frac{\varepsilon}{3} \tag{11.9.4}$$

for every $\sigma > 0$. We remark that for every $k, \sigma > 0$ and for every $n, m \in \mathbb{N}$, one has

$$\begin{aligned} \{|u_n - u_m| > \sigma\} &\subseteq [\{|u_n| > k\} \cap \{|u_n - u_m| > \sigma\}] \\ &\quad \cup [\{|u_n| \leq k\} \cap \{|u_m| > k\} \cap \{|u_n - u_m| > \sigma\}] \\ &\quad \cup [\{|u_n| \leq k\} \cap \{|u_m| \leq k\} \cap \{|u_n - u_m| > \sigma\}] \\ &\subseteq \{|u_n| > k\} \cup \{|u_m| > k\} \cup \{|T_k(u_n) - T_k(u_m)| > \sigma\} . \end{aligned}$$

By estimates (11.9.3) and (11.9.4) we deduce that, for $k > \overline{k}$ and for every $n, m > \eta_\varepsilon$

$$\{|u_n - u_m| > \sigma\} \leq \varepsilon .$$

This implies that u_n is a Cauchy sequence in measure. Therefore there exists a function u such that u_n converges to u in measure and consequently, for every $k > 0$, $T_k(u_n) \to T_k(u)$ in measure. By uniqueness of the limit $T_k(u) = v_k$. Since $T_k(u_n) \to v_k$ weakly in $H_0^1(\Omega)$ we deduce that $T_k(u)$ belongs to $H_0^1(\Omega)$ for every k. □

Theorem 11.28. *Let $f \in L^1(\Omega)$. Then there exists a T-minimum u of J such that, for every $k, h > 0$,*

$$\int_\Omega |\nabla T_k(u)|^2 \leq \frac{\|f\|_{L^1(\Omega)}}{\alpha} k; \tag{11.9.5}$$

$$\int_{\{h \leq |u| \leq h+k\}} |\nabla u|^2 \, dx \leq \frac{k}{\alpha} \int_{\{|u| \geq h\}} |f| . \tag{11.9.6}$$

Proof. Let $f_n = T_n(f)$. Then the functions

$$v \to \int_\Omega j(x, v) - \int_\Omega f_n v$$

have minimizers $u_n \in H_0^1(\Omega) \cap L^\infty(\Omega)$ by Theorem 9.15. Then u_n satisfies

$$\int_\Omega j(x, \nabla u_n) - \int_\Omega f_n u_n \leq \int_\Omega j(x, \nabla(u_n - T_k(u_n))) - \int_\Omega f_n(u_n - T_k(u_n))$$

for every $k > 0$. By (11.9.1), we have

$$\alpha \int_\Omega |\nabla T_k(u_n)|^2 \leq \int_\Omega f_n T_k(u_n) \leq \|f\|_{L^1(\Omega)} k .$$

For any fixed $k > 0$, the sequence $T_k(u_n)$ is bounded in $H_0^1(\Omega)$. Therefore, by Lemma 11.27 there exist a subsequence (not relabeled) and a function $u \in \mathcal{T}(\Omega)$ such that u_n converges to u a.e. in Ω and, for every $k > 0$, $T_k(u_n)$ converges to $T_k(u)$ weakly in $H_0^1(\Omega)$.

Let $\varphi \in H_0^1(\Omega) \cap L^\infty(\Omega)$; again the minimality of u_n yields:

$$\int_\Omega j(x, \nabla u_n) \le \int_\Omega j(x, \nabla(u_n - T_k(u_n - \varphi))) + \int_\Omega f_n T_k(u_n - \varphi).$$

This is equivalent to

$$\int_{\{|u_n-\varphi|\le k\}} j(x, \nabla(\varphi + T_k(u_n - \varphi))) + \int_{\{|u_n-\varphi|>k\}} j(x, \nabla u_n)$$
$$\le \int_{\{|u_n-\varphi|\le k\}} j(x, \nabla\varphi) + \int_{\{|u_n-\varphi|>k\}} j(x, \nabla u_n) + \int_\Omega f_n T_k(u_n - \varphi)$$

which finally implies

$$\int_\Omega j(x, \nabla(\varphi + T_k(u_n - \varphi))) \le \int_\Omega j(x, \nabla\varphi) + \int_\Omega f_n T_k(u_n - \varphi). \tag{11.9.7}$$

The weak $H_0^1(\Omega)$ lower semicontinuity of $v \to \int_\Omega j(x, \nabla v)$ (assured by Theorem 9.2) and the weak $H_0^1(\Omega)$ convergence of $T_k(u_n - \varphi)$ to $T_k(u - \varphi)$ allow us to pass to the limit in (11.9.7) and to obtain the existence of a T-minimum u, according to Definition 11.26.

Assumption (11.9.1) and the choice of $\varphi = 0$ in (11.9.7) give (11.9.5). As well, using $\varphi = T_h(u)$ in (11.9.7), we easily get (11.9.6). □

We assume in the sequel that $j(x, \xi)$ is differentiable with respect to ξ and $a(x, \xi) = \nabla_\xi j(x, \xi)$ satisfies

(1) there exists $\beta > 0$ such that $|a(x, \xi)| \le \beta|\xi|$;
(2) there exists $\alpha > 0$ such that $a(x, \xi) \cdot \xi \ge \alpha|\xi|^2, \quad \forall\, \xi \in \mathbb{R}^N$;
(3) $[a(x, \xi) - a(x, \eta)] \cdot [\xi - \eta] > 0$ if $\xi \ne \eta$.

We show the strict relationship between T-minima and entropy solutions to the boundary value problem

$$\begin{cases} -\operatorname{div}(a(x, \nabla u)) = f, & \text{in } \Omega, \\ u = 0, & \text{on } \partial\Omega. \end{cases} \tag{11.9.8}$$

Proposition 11.29. *Let $f \in L^1(\Omega)$. Under the above assumptions, let u be an entropy solution to* (11.9.8), *that is,*

$$\int_\Omega \nabla_\xi(x, \nabla u) \nabla T_k(u - \varphi) \le \int_\Omega f T_k(u - \varphi),$$

$\forall\, \varphi \in H_0^1(\Omega) \cap L^\infty(\Omega)$. *Then u is a T-minimum of J.*

Proof. The convexity of $j(x,\xi)$ with respect to ξ implies that

$$j(x,\nabla\varphi) \geq j(x,\nabla u) + \nabla_\xi j(x,\nabla u)\cdot(\nabla\varphi - \nabla u)\,.$$

The integration of the previous inequality on the set $\{|u-\varphi| \leq k\}$ yields the conclusion. □

Theorem 11.30. *Let $f \in L^1(\Omega)$. Under the previous assumptions, let u be a T-minimum of J. Then u is an entropy solution to the boundary value problem* (11.9.8).

Proof. It is clear that

$$\int\limits_{\substack{\{|u-\varphi|\leq k\}\\ \cap\{|u|\leq h\}}} j(x,\nabla u) \leq \int\limits_{\substack{\{|u-\varphi|\leq k\}\\ \cap\{|u|\leq h\}}} j(x,\nabla u) + \int\limits_{\substack{\{|u-\varphi|\leq k\}\\ \cap\{h<|u|\}}} j(x,\nabla u) = \int\limits_{\{|u-\varphi|\leq k\}} j(x,\nabla u)\,.$$

In the definition of T-minimum, we consider $\varphi = T_h(u) + tT_k(v - T_h(u))$, with $0 < t < 1$ and $v \in H_0^1(\Omega)\cap L^\infty(\Omega)$. We thus have

$$\begin{aligned}\int\limits_{\{|u-\varphi|\leq k\}} j(x,\nabla u) \leq & \int\limits_{\substack{\{|u-\varphi|\leq k\}\cap\{|u-v|\leq k\}\\ \cap\{|u|\leq h\}}} j(x,\nabla u + t\nabla(v-u)) \\ & + \int\limits_{\substack{\{|u-\varphi|\leq k\}\cap\{k<|u-v|\}\\ \cap\{|u|\leq h\}}} j(x,\nabla u) \\ & + \int\limits_{\substack{\{|u-\varphi|\leq k\}\cap\{|v-T_h(u)|\leq k\}\\ \cap\{h<|u|\}}} j(x,t\nabla v) \\ & + \int\limits_{\Omega} fT_k(u - T_h(u) - tT_k(v - T_h(u)))\,.\end{aligned}$$

We can estimate the difference between the first and the third integral in this way:

$$\begin{aligned}\int\limits_{\{|u-v|\leq k\}\cap\{|u|\leq h\}} j(x,\nabla u) \leq & \int\limits_{\{|u-v|\leq k\}\cap\{|u|\leq h\}} j(x,\nabla u + t\nabla(v-u)) \\ & + \int\limits_{\substack{\{|u-\varphi|\leq k\}\cap\{<|v-T_h(u)|\leq k\}\\ \cap\{h<|u|\}}} j(x,t\nabla v) \\ & + \int\limits_{\Omega} fT_k(u - T_h(u) - tT_k(v - T_h(u)))\,,\end{aligned}$$

where we have used that $\{|u| \leq h\} \subset \{|u-\varphi| \leq k\}$. Since $T_j(u) \in H_0^1(\Omega)$, for any $j > 0$, and $v \in H_0^1(\Omega)$, we can pass to the limit as $h \to \infty$, by Lebesgue theorem. Moreover, since $0 < t < 1$, $T_k(tT_k(v-u)) = tT_k(v-u)$. Thus

$$-\int\limits_{\{|u-v|\leq k\}} \frac{j(x,\nabla u + t\nabla(v-u)) - j(x,\nabla u)}{t} \leq \int\limits_{\Omega} fT_k(u-v)\,.$$

The conclusion follows passing to the limit as $t \to 0^+$. □

11.10 Appendix

The aim of this section is to recall the definition of $\mathcal{M}(\Omega)$ that we have used in Section 11.7 (see [29] for further details).

We denote by $\mathcal{P}(\Omega)$ the set of subsets of Ω such that $\mathcal{P}(\Omega) = \{E : E \subseteq \Omega\}$.

Definition 11.31. A σ-algebra of sets over Ω is a family of sets $\Sigma \subset \mathcal{P}(\Omega)$ such that
(1) $\emptyset \in \Sigma$;
(2) if $E \in \Sigma$ then $CE \in \Sigma$, where CE is $\Omega \backslash E$;
(3) for every sequence $\{E_k\} \subseteq \Sigma$, $\bigcup_{k=1}^{+\infty} E_k \in \Sigma$.

(Ω, Σ) is called measurable space.

Definition 11.32. The σ-algebra of the Borel subsets of Ω, $B(\Omega)$, is the smallest σ-algebra which contains the open subsets of Ω.

Definition 11.33. Let (Ω, Σ) be a measurable space. A signed measure μ over Σ is a map from Σ to $[-\infty, +\infty]$ such that
(1) $\mu(\emptyset) = 0$;
(2) $\mu\left(\bigcup_{k=1}^{+\infty} E_k\right) = \sum_{k=1}^{\infty} \mu(E_k)$ for every sequence $\{E_k\} \subseteq \Sigma$ of disjoint sets;
(3) it does not take the values $-\infty$ and $+\infty$ both.

Definition 11.34. Let (Ω, Σ) be a measurable space. A measure μ over Σ is called a Borel measure on Ω if $B(\Omega) \subseteq \Sigma$.

Definition 11.35. Let (Ω, Σ) be a measurable space. A Borel measure μ is called regular if, for every $\delta > 0$ and for every $E \in \Sigma$, there exist a compact set $K_\delta \subseteq E$ and an open set $A_\delta \supseteq E$, such that $\mu(E \setminus K_\delta) \leq \delta$ and $\mu(A_\delta \setminus E) \leq \delta$.

Definition 11.36. Let (Ω, Σ) be a measurable space. Let μ be a signed measure over Σ.
(1) The positive variation of μ is $\mu^+(E) = \sup\{\mu(F) : F \in \Sigma, F \subseteq E\}$, for every set $E \in \Sigma$.
(2) The negative variation of μ is $\mu^-(E) = (-\mu)^+(E)$, for every set $E \in \Sigma$.
(3) The total variation of μ is the measure $|\mu| = \mu^+ + \mu^-$.

Definition 11.37. $\mathcal{M}(\Omega)$ is the set of regular signed measures with finite total variation on the σ-algebra of the Borel sets of Ω. On $\mathcal{M}(\Omega)$, $\|\mu\|_{\mathcal{M}(\Omega)} = |\mu|(\Omega)$ is a norm.

Definition 11.38. If μ_n is a sequence of measures in $\mathcal{M}(\Omega)$, μ_n converges $*$-weakly to μ if

$$\lim_{n\to+\infty} \int_\Omega \varphi \, d\mu_n = \int_\Omega \varphi \, d\mu,$$

for every continuous function φ on $C_0(\Omega)$.

Theorem 11.39. *If $\mu \in \mathcal{M}(\Omega)$, then there exists a sequence f_n of $C_0^\infty(\Omega)$ functions such that*

(1) $\|f_n\|_{L^1(\Omega)} \le \|\mu\|_{\mathcal{M}(\Omega)}$, $\forall n \in \mathbb{N}$;

(2) $f_n \to \mu$ *$*$-weakly in $\mathcal{M}(\Omega)$.*

Definition 11.40. Let f_n be a sequence functions and $\mu \in \mathcal{M}(\Omega)$. We say that f_n converges to μ in the tight convergence of measures, if

$$\lim_{n\to+\infty} \int_\Omega f_n \varphi = \int_\Omega \varphi \, d\mu, \qquad \forall \varphi \in C(\overline{\Omega}).$$

12 Uniqueness

12.1 Introduction

In Chapters 5 and 11, we have seen a result of existence of solutions to problem

$$\begin{cases} -\operatorname{div}(a(x,u,\nabla u)) = f\,, & \text{in} \quad \Omega\,, \\ u = 0\,, & \text{on} \quad \partial\Omega\,, \end{cases}$$

with $f \in L^m(\Omega)$, for $m \geq 1$. In this chapter we give some uniqueness results. We have to say that, in contrast with the existence problem, for the uniqueness question, unfortunately, there are only some partial results, and this is independent from the summability of the source f. The uniqueness of solutions is more related to the monotonicity of $A(v) = -\operatorname{div}(a(x,v,\nabla v))$ than to the summability of the source.

In the first section of this chapter, we give a general result for the problem

$$\begin{cases} -\operatorname{div}(a(x,\nabla u)) = f\,, & \text{in} \quad \Omega\,, \\ u = 0\,, & \text{on} \quad \partial\Omega\,, \end{cases}$$

that is, for a problem defined through a monotone operator. We will not present the original proof in [4], but that one in [43], which is a slightly simpler. In the second part we present a uniqueness result for a particular nonmonotone operator (following [3] and [15]). Finally, we will treat the case of measure sources for linear problems.

In the whole chapter Ω denotes an open bounded subset of $\mathbb{R}^N$, $N \geq 3$.

12.2 Monotone elliptic operators

In this section, we study the uniqueness of solutions to

$$\begin{cases} -\operatorname{div}(a(x,\nabla u)) = f\,, & \text{in} \quad \Omega\,, \\ u = 0\,, & \text{on} \quad \partial\Omega\,, \end{cases} \tag{12.2.1}$$

under the following assumptions: $a : \Omega \times \mathbb{R}^N \to \mathbb{R}^N$ is a Carathéodory map, such that

(1) there exists $\beta > 0$ such that $|a(x,\xi)| \leq \beta|\xi|$;
(2) there exists $\alpha > 0$ such that $a(x,\xi)\cdot\xi \geq \alpha|\xi|^2$;
(3) $[a(x,\xi) - a(x,\eta)]\cdot[\xi-\eta] > 0$ if $\xi \neq \eta$.

Definition 12.1. Let V be a Banach reflexive space. An operator $A : V \to V'$ is monotone if $\langle A(u) - A(v), u - v\rangle \geq 0$ for all $u, v \in V$.

Observe that assumption 3 on a implies that $-\operatorname{div}(a(x,\nabla u))$ is a monotone operator.

We will prove the uniqueness of solutions among the entropy solutions.

Theorem 12.2. *Let f in $L^1(\Omega)$. Then there exists a unique entropy solution to problem* (12.2.1).

Proof. Let u be a solution obtained by approximation as in Theorem 11.4. It is sufficient to prove that any other entropy solution is equal to u. We briefly recall how we obtained u. Let $u_n \in H_0^1(\Omega)$ be the solutions to

$$-\operatorname{div}(a(x, \nabla u_n)) = f_n \quad \text{in} \quad \Omega, \tag{12.2.2}$$

where $f_n = T_n(f)$; $u_n \in L^\infty(\Omega)$ for every n from Theorem 6.6. Moreover, $u_n(x)$ and $\nabla u_n(x)$ converge, respectively, to $u(x)$ and $\nabla u(x)$, a.e. in Ω (see Lemmata 11.1 and 11.3). We recall that u is an entropy solution. Let z be another entropy solution, that is, $T_k(z) \in H_0^1(\Omega)$ for every k and z satisfies

$$\int_\Omega a(x, \nabla z) \cdot \nabla T_k(z - \varphi) \le \int_\Omega f T_k(z - \varphi), \tag{12.2.3}$$

for every $\varphi \in H_0^1(\Omega) \cap L^\infty(\Omega)$. Let us choose $\varphi = u_n$ in (12.2.3). We get

$$\int_\Omega a(x, \nabla z) \cdot \nabla T_k(z - u_n) \le \int_\Omega f T_k(z - u_n). \tag{12.2.4}$$

On the other hand, using $T_k(z - u_n)$ as a test function in (12.2.2), one has

$$-\int_\Omega a(x, \nabla u_n) \cdot \nabla T_k(z - u_n) = -\int_\Omega f_n T_k(z - u_n). \tag{12.2.5}$$

Adding up (12.2.4) and (12.2.5) gives

$$\int_\Omega [a(x, \nabla z) - a(x, \nabla u_n)] \cdot \nabla T_k(z - u_n) \le \int_\Omega (f - f_n) T_k(z - u_n).$$

The integrand of the left-hand side of the previous inequality is positive from the monotonicity of a. Moreover $[a(x, \nabla z) - a(x, \nabla u_n)] \cdot \nabla T_k(z - u_n)$ converges a.e. in Ω to $[a(x, \nabla z) - a(x, \nabla u)] \cdot \nabla T_k(z - u)$. The right-hand side goes to 0, for every fixed k, by Lebesgue's theorem. Using Fatou's lemma, we can pass to the limit for $n \to +\infty$ getting

$$\int_\Omega [a(x, \nabla z) - a(x, \nabla u)] \cdot \nabla T_k(z - u) \le 0.$$

The monotonicity of a implies $T_k(z - u) = 0$ a.e. in Ω for every $k > 0$. Therefore, $z = u$ a.e. in Ω. □

In the case where $f \in L^{\frac{2N}{N+2}}(\Omega)$, from Chapters 5 and 11 we know that there exists a $H_0^1(\Omega)$ solution, which is also an entropy solution. One can prove the uniqueness of solutions in a more simple way as the following theorem shows.

Theorem 12.3. *Let* $f \in L^{\frac{2N}{N+2}}(\Omega)$. *Then problem*

$$\begin{cases} -\operatorname{div}(a(x, \nabla u)) = f, & \text{in } \Omega, \\ u = 0, & \text{on } \partial\Omega, \end{cases}$$

has a unique solution $u \in H_0^1(\Omega)$.

Proof. Let u_1 and $u_2 \in H_0^1(\Omega)$ be two solutions. Let us use $u_1 - u_2$ as a test function:

$$\int_\Omega a(x, \nabla u_1) \cdot (\nabla u_1 - \nabla u_2) = \int_\Omega f(u_1 - u_2)$$

$$\int_\Omega a(x, \nabla u_2) \cdot (\nabla u_1 - \nabla u_2) = \int_\Omega f(u_1 - u_2).$$

Subtracting side by side and using assumption 3 on a we get $u_1 = u_2$ a.e. in Ω. □

The following example shows that even a linear problem may have more than one distributional solution.

Remark 12.4. Let M, v and f be defined by (11.6.1), (11.6.4), and (11.6.3) respectively. The Dirichlet problem

$$\begin{cases} -\operatorname{div}(M(x)\nabla v) = f, & \text{in } \Omega, \\ v = 0, & \text{on } \partial\Omega, \end{cases}$$

where $\Omega = \{x \in \mathbb{R}^N : |x| < 1\}$ has two distributional solutions. Indeed v is a solution, as seen in Section 11.6; moreover there exists a solution v_1 in $H_0^1(\Omega)$, since $f \in L^\infty(\Omega)$.

By linearity one can give the following striking example. Let $w = v - v_1$: note that $w \in W_0^{1,q}(\Omega)$, $q < \frac{N}{N-1}$. By linearity, w solves

$$\begin{cases} -\operatorname{div}(M(x)\nabla w) = 0, & \text{in } \Omega, \\ w = 0, & \text{on } \partial\Omega. \end{cases}$$

It is clear that 0 is a solution to this problem. Therefore, the previous problem has two solutions: the zero solution (entropy solution) and the distributional solution w (which is not an entropy solution).

12.3 A nonmonotone elliptic operator

In this section, we will give a uniqueness result for

$$\begin{cases} -\operatorname{div}(M(x, u)\nabla u) = f, & \text{in } \Omega, \\ u = 0, & \text{on } \partial\Omega. \end{cases} \tag{12.3.1}$$

We observe that $A(v) = -\mathrm{div}(M(x,v)\,\nabla v)$ is not monotone (see [9] for more details). We will prove the following theorem:

Theorem 12.5. *Let $M(x,s)$ be a $N \times N$ matrix composed by Carathéodory, bounded and Lipschitz continuous functions in the second variable. We assume that there exists $\alpha > 0$ such that*

$$M(x,s)\xi \cdot \xi \geq \alpha\,|\xi|^2 .$$

Then for every $f \in L^{\frac{2N}{N+2}}(\Omega)$ there exists a weak solution $u \in H_0^1(\Omega)$ to problem (12.3.1).

Proof. The existence of solutions follows from Leray–Lions theorem (Theorem 5.1). Now, let u_1 and u_2 be two solutions, that is, u_1 and u_2 satisfy

$$\int_\Omega M(x,u_1)\nabla u_1 \cdot \nabla v = \int_\Omega f\,v$$

and

$$\int_\Omega M(x,u_2)\nabla u_2 \cdot \nabla v = \int_\Omega f\,v\,,$$

for every $v \in H_0^1(\Omega)$. We then get

$$\int_\Omega M(x,u_1)\nabla(u_1-u_2)\cdot\nabla v = \int_\Omega [M(x,u_2)-M(x,u_1)]\,\nabla u_2\cdot\nabla v\,. \qquad (12.3.2)$$

Let us fix b and B such that $0 < b < B$. Since m_{ij} is Lipschitz continuous, there exists $L > 0$ such that

$$|M(x,u_1)-M(x,u_2)| \leq L\,|u_1-u_2|\,.$$

The use of $v = \frac{u_1-u_2}{b+|u_1-u_2|}$ as a test function in (12.3.2) implies

$$\int_\Omega M(x,u_1)\nabla(u_1-u_2)\cdot\nabla(u_1-u_2)\frac{1}{(b+|u_1-u_2|)^2}$$
$$= \int_\Omega [M(x,u_2)-M(x,u_1)]\nabla u_2\cdot\nabla(u_1-u_2)\frac{1}{(b+|u_1-u_2|)^2}\,.$$

Consequently, the ellipticity of M and the fact that m_{ij} is Lipschitz continuous give

$$\alpha\int_\Omega \frac{|\nabla(u_1-u_2)|^2}{[b+|u_1-u_2|]^2} \leq \int_\Omega \frac{L|\nabla u_2||\nabla(u_1-u_2)|}{b+|u_1-u_2|}\,.$$

Hölder's inequality on the right-hand side implies

$$\alpha\left(\int_\Omega \left|\nabla \log\left(1+\frac{|u_1-u_2|}{b}\right)\right|^2\right)^{1/2} \leq L\,\|u_2\|_{H_0^1(\Omega)}\,.$$

By using Poincaré's inequality on the left-hand side we get

$$\alpha c \left(\int_\Omega \left| \log \left(1 + \frac{|u_1 - u_2|}{b} \right) \right|^2 \right)^{1/2} \le L \, \|u_2\|_{H_0^1(\Omega)} .$$

Therefore

$$\alpha c \operatorname{meas}(\{|u_1 - u_2| > B\})^{\frac{1}{2}} \left| \log \left(1 + \frac{B}{b} \right) \right| \le L \, \|u_2\|_{H_0^1(\Omega)} .$$

If $b \to 0^+$, necessarily $\operatorname{meas}(\{|u_1 - u_2| > B\}) = 0$ for every $B > 0$. Therefore $u_1 = u_2$ a.e. in Ω. □

12.4 A uniqueness result for measure sources

In this section, we study the elliptic problem

$$\begin{cases} -\operatorname{div}(M(x)\nabla u) = \mu , & \text{in } \Omega , \\ u = 0 , & \text{on } \partial\Omega , \end{cases} \tag{12.4.1}$$

where $\mu \in \mathcal{M}(\Omega)$ and $M \in \mathbb{R}^{N \times N}$ is a symmetric matrix with $L^\infty(\Omega)$ entries, such that $M(x)\xi \cdot \xi \ge \alpha|\xi|^2$. We assume that $\partial\Omega \in C^1$. Theorem 11.4 guarantees the existence of a solution. We recall that we obtained this solution by approximation in Theorem 11.19. In this section we give a uniqueness result of solutions as limit of the solutions to

$$\begin{cases} -\operatorname{div}(M(x)\nabla u_n) = f_n , & \text{in } \Omega , \\ u_n = 0 , & \text{on } \partial\Omega , \end{cases} \tag{12.4.2}$$

where f_n is any sequence of regular functions that converges weakly (in the sense of measures) to μ and which is bounded in $L^1(\Omega)$.

Theorem 12.6. *Under the above hypotheses, let u and v be two solutions to* (12.4.1) *obtained by approximation from problems* (12.4.2). *Then $u = v$ a.e. in Ω.*

The following lemma will be useful to us:

Lemma 12.7. *Let $f_n : \Omega \to \mathbb{R}$ be a sequence of functions such that $\int_\Omega \frac{f_n^2}{1+|f_n|} \to 0$. Then $f_n \to 0$ in $L^1(\Omega)$.*

Proof. We have

$$\int_\Omega \frac{f_n^2}{1 + |f_n|} \ge \int_{\{|f_n|>1\}} \frac{f_n^2}{1 + |f_n|} \ge \frac{1}{2} \int_{\{|f_n|>1\}} |f_n| .$$

For every fixed $t > 0$ one has

$$\int_\Omega \frac{f_n^2}{1 + |f_n|} \ge \int_{\{t<|f_n|<1\}} \frac{f_n^2}{1 + |f_n|} \ge \frac{t^2}{1+t} \operatorname{meas}(\{t < |f_n| < 1\}) .$$

Consequently

$$\begin{aligned}\int_\Omega |f_n| &= \int_{\{|f_n|\ge 1\}} |f_n| + \int_{\{t\le|f_n|\le 1\}} |f_n| + \int_{\{|f_n|\le t\}} |f_n| \\ &\le 2\int_\Omega \frac{f_n^2}{1+|f_n|} + \operatorname{meas}(\{t\le |f_n|\le 1\}) + t\operatorname{meas}(\Omega) \\ &\le 2\int_\Omega \frac{f_n^2}{1+|f_n|} + \frac{1+t}{t^2}\int_\Omega \frac{f_n^2}{1+|f_n|} + t\operatorname{meas}(\Omega)\,.\end{aligned}$$

Passing to the limit as $n \to \infty$, the previous inequality gives

$$\lim_{n\to\infty}\int_\Omega |f_n| \le t\operatorname{meas}(\Omega)\,, \quad \forall\, t > 0\,.$$

Since t is arbitrary, $f_n \to 0$ in $L^1(\Omega)$. □

The following theorem will be useful too (for the proof use Theorem 8.29 in [30], due to De Giorgi, Theorems 6.6 and 6.14):

Theorem 12.8. *Under the previous hypotheses on M and Ω, let $F \in (L^q(\Omega))^N, q > N$, and $f \in (L^s(\Omega))^N, s > \frac{N}{2}$. Then the solution $u \in H_0^1(\Omega)$ to*

$$\begin{cases} -div(M(x)\nabla u) = divF + f\,, & \text{in} \quad \Omega\,, \\ u = 0\,, & \text{on} \quad \partial\Omega\,, \end{cases}$$

is Hölder continuous in $\overline{\Omega}$ and satisfies

$$\sup_{x,y\in\overline{\Omega}} \frac{|u(x)-u(y)|}{|x-y|^\beta} \le C[\|F\|_{L^q(\Omega)} + \|f\|_{L^s(\Omega)}]$$

where $C = C(\Omega, \alpha, N, q, s)$ and $\beta = \beta(\Omega, \alpha, N, q, s)$.

We can now prove Theorem 12.6.

Proof. Let u and v be two solutions obtained by approximation from problems (12.4.2). Then there exist two sequences of regular functions, f_n and g_n, converging weakly in the sense of measures to μ and f_n and g_n are bounded in $L^1(\Omega)$. Moreover, if u_n and v_n are the $H_0^1(\Omega)$ solutions to the approximating problems (12.4.2), u_n and v_n converge weakly to u and v, respectively, in $W_0^{1,q}(\Omega)$, $\forall\, q < \frac{N}{N-1}$, by Lemma 11.1. By linearity we have that

$$\int_\Omega M(x)\nabla(u_n - v_n)\cdot\nabla\varphi = \int_\Omega (f_n - g_n)\varphi\,, \quad \forall\, \varphi \in H_0^1(\Omega)\,. \tag{12.4.3}$$

On the other hand, let $z_n \in H_0^1(\Omega)$ be a solution to

$$-\operatorname{div}(M(x)\nabla z_n) = -\operatorname{div}\left(\frac{\nabla(u_n - v_n)}{1 + |\nabla(u_n - v_n)|}\right)\,. \tag{12.4.4}$$

$F_n = \frac{\nabla(u_n - v_n)}{1+|\nabla(u_n-v_n)|}$ belongs to $(L^\infty(\Omega))^N$ and $\mathrm{div}F_n \in L^2(\Omega)$. The existence of $z_n \in H_0^1(\Omega)$ is guaranteed for every n by Theorem 5.1. We observe that z_n are uniformly Hölder and uniformly bounded in Ω by Theorem 12.8. The Arzelà–Ascoli theorem implies that, up to subsequence, $z_n \to z$ uniformly in $\overline{\Omega}$ for some continuous function z in $\overline{\Omega}$. We have, choosing $u_n - v_n$ as a test function in (12.4.4)

$$\int_\Omega \frac{|\nabla(u_n - v_n)|^2}{1+|\nabla(u_n - v_n)|} = \int_\Omega M(x)\nabla z_n \cdot \nabla(u_n - v_n)$$
$$= \int_\Omega M(x)\nabla(u_n - v_n)\cdot \nabla z_n = \int_\Omega (f_n - g_n)\, z_n$$

by (12.4.3). Writing $(f_n - g_n)z_n$ as $(f_n - g_n)(z_n - z) + (f_n - g_n)\,z$, it is easy to prove that the last integral goes to 0, since $z_n \to z$ uniformly in Ω, $f_n - g_n$ goes to 0 in the sense of measures and is bounded in $L^1(\Omega)$. Therefore

$$\int_\Omega \frac{|\nabla(u_n - v_n)|^2}{1+|\nabla(u_n - v_n)|} \to 0\,, \quad n \to \infty\,.$$

By Lemma 12.7, this implies that $\nabla(u_n - v_n) \to 0$ in $L^1(\Omega)$. Since u_n and v_n converge weakly to u and v in $W_0^{1,q}(\Omega)$, for every $q < \frac{N}{N-1}$, one has $\nabla(u - v) = 0$ and so $u = v$ a.e. in Ω. □

Remark 12.9. We are going to prove that it is not possible to find a solution to the Dirichlet problem $u \in W_0^{1,p}(\Omega)$: $-\mathrm{div}(M(x)\nabla u) = -\mathrm{div}F$ for every $F \in (L^p(\Omega))^N$, $p > N$, where M is defined by (11.6.1). Assume, by contradiction that there exists a solution u, that is,

$$\int_\Omega M(x)\nabla u \cdot \nabla v = \int_\Omega F\cdot\nabla v, \qquad \forall\, v \in W_0^{1,p'}(\Omega)\,.$$

In particular one can choose, as a test function, the not null function $w \in W_0^{1,q}(\Omega)$, $q < \frac{N}{N-1}$, defined in Remark 12.4, and $F = |\nabla w|^{q-2}\nabla w$, so that

$$\int_\Omega M(x)\nabla u \cdot \nabla w = \int_\Omega |\nabla w|^q\,.$$

On the other hand, one can choose u as a test function in the weak formulation of the problem solved by w, that is,

$$\int_\Omega M(x)\nabla w \cdot \nabla u = 0\,.$$

By the symmetry of M, the last two equalities give $w = 0$, which is a contradiction.

13 A problem with polynomial growth

13.1 Introduction

In this chapter, we will study a polynomial growth elliptic equation. More precisely, we will prove existence and regularity results for the solutions to

$$\begin{cases} -\operatorname{div}(a(x,u,\nabla u)) = f\,, & \text{in} \quad \Omega\,, \\ u = 0\,, & \text{on} \quad \partial\Omega\,, \end{cases} \tag{13.1.1}$$

under the following assumptions. The set Ω is an open bounded subset of $\mathbb{R}^N$, $N > 2$, and $a : \Omega \times \mathbb{R} \times \mathbb{R}^N \to \mathbb{R}^N$ is a Carathéodory map, with the following properties:

(1) there exists $\beta > 0$ such that $|a(x,s,\xi)| \le \beta\, b(|s|)|\xi|$;

(2) $a(x,s,\xi)\cdot\xi \ge b(|s|)|\xi|^2, \quad \forall\, \xi \in \mathbb{R}^N$;

(3) $[a(x,s,\xi) - a(x,s,\eta)]\cdot[\xi-\eta] > 0$ if $\xi \ne \eta$

where $b : [0,+\infty) \to (0,+\infty)$ is a continuous function such that

$$b(|s|) \ge \gamma\,|s|^r\,, \quad \forall\,|s| \ge s_0\,, \tag{13.1.2}$$

$\gamma > 0$, $r \ge 0$ and $s_0 > 0$. Moreover, there exists $\alpha > 0$ such that for every $s \in \mathbb{R}$

$$b(|s|) \ge \alpha\,. \tag{13.1.3}$$

The polynomial growth of a will give us more regular solutions than the solutions found in Chapters 6 and 11 for the Leray–Lions problem. We will prove the existence of distributional solutions u, that is,

$$\int_\Omega a(x,u,\nabla u)\cdot\nabla\varphi = \int_\Omega f\varphi\,, \quad \forall\,\varphi \in C_0^\infty(\Omega)\,.$$

We describe in a schematic way the existence results of distributional solutions u that we are going to prove (following [42]).

(1) Let $f \in L^1(\Omega)$.
   (a) If $r > 1$, then u belongs to $H_0^1(\Omega) \cap L^s(\Omega)$, $s < \frac{N(r+1)}{N-2}$.
   (b) If $0 \le r \le 1$, then u belongs to $W_0^{1,q}(\Omega)$, $q < \frac{N(r+1)}{N-1+r}$.

(2) Let $f \in L^m(\Omega)$, $1 < m < \frac{2N}{N+2}$.
   (a) If $r \ge 1 - \frac{2^*}{m'}$, then u belongs to $H_0^1(\Omega) \cap L^{\frac{Nm(r+1)}{N-2m}}(\Omega)$.
   (b) If $0 \le r < 1 - \frac{2^*}{m'}$, then u belongs to $W_0^{1,\frac{Nm(r+1)}{N-m(1-r)}}(\Omega)$.

(3) Let $f \in L^m(\Omega)$, $m \ge \frac{2N}{N+2}$.
   (a) If $m > \frac{N}{2}$, then u belongs to $H_0^1(\Omega) \cap L^\infty(\Omega)$.
   (b) If $\frac{2N}{N+2} \le m < \frac{N}{2}$, then u belongs to $H_0^1(\Omega) \cap L^{\frac{Nm(r+1)}{N-2m}}(\Omega)$.

13.2 Existence results

We define the following approximating problems:

$$\begin{cases} -\mathrm{div}(a(x, T_n(u_n), \nabla u_n)) = f_n\,, & \text{in } \Omega\,, \\ u_n = 0\,, & \text{on } \partial\Omega\,, \end{cases} \tag{13.2.1}$$

where $f_n = T_n(f)$. The existence of a bounded $H_0^1(\Omega)$ is assured by Theorems 5.1 and 6.6.

Lemma 13.1. *Let u_n be a $H_0^1(\Omega) \cap L^\infty(\Omega)$ solution to problem* (13.2.1). *Then*

(1) *for every $k > 0$, $\|T_k(u_n)\|^2_{H_0^1(\Omega)} \le k \frac{\|f\|_{L^1(\Omega)}}{\alpha}$;*

(2) *$|\nabla B(u_n)|$ is bounded in $M^{\frac{N}{N-1}}(\Omega)$, where $B(t) = \int_0^t b(|s|)$;*

(3) *$a(x, T_n(u_n), \nabla u_n)$ is bounded in $L^q(\Omega)$ for every $q < \frac{N}{N-1}$.*

Proof. (1) Taking $T_k(u_n)$ as a test function in (13.2.1) and using (13.1.3) we get that

$$\|T_k(u_n)\|^2_{H_0^1(\Omega)} \le k \frac{\|f\|_{L^1(\Omega)}}{\alpha}\,.$$

(2) Let $B(t) = \int_0^t b(|s|)$; it is easy to see that $T_k(B(u_n))$ belongs to $H_0^1(\Omega)$, so that it can be taken as a test function in (13.2.1). The assumptions on a give

$$\int_{\{|B(u_n)| \le k\}} b(|u_n|)^2 |\nabla u_n|^2 \le \int_{\{|B(u_n)| \le k\}} a(x, T_n(u_n), \nabla u_n) \cdot \nabla u_n b(|u_n|) \le k\|f\|_{L^1(\Omega)}\,,$$

and then the following estimate holds:

$$\int_\Omega |\nabla T_k(B(u_n))|^2 \le k\,\|f\|_{L^1(\Omega)}\,. \tag{13.2.2}$$

Proposition 11.13 implies that $|\nabla B(u_n)|$ is bounded in $M^{\frac{N}{N-1}}(\Omega)$, that is, $b(|u_n|)|\nabla u_n|$ is bounded in $M^{\frac{N}{N-1}}(\Omega)$.

(3) Since $\{|a(x, T_n(u_n), \nabla u_n)| > k\}$ is contained in $\{\beta b(|u_n|)|\nabla u_n| > k\}$, we conclude that $a(x, T_n(u_n), \nabla u_n)$ is bounded in $L^q(\Omega)$ for every $q < \frac{N}{N-1}$. ☐

In the sequel C denotes a positive constant independent of $n \in \mathbb{N}$.

Lemma 13.2. *Let $f \in L^1(\Omega)$.*

(1) *If $r > 1$, then the sequence of the solutions u_n to problem* (13.2.1) *is bounded in $H_0^1(\Omega) \cap L^s(\Omega)$, $s < \frac{N(r+1)}{N-2}$.*

(2) *If $0 \le r \le 1$, then the sequence of the solutions u_n to problem* (13.2.1) *is bounded in $W_0^{1,q}(\Omega)$, $q < \frac{N(r+1)}{N-1+r}$.*

Proof. Let u_n be a solution to problem (13.2.1). We consider $T_1(u_n - T_k(u_n))$, with $k \ge s_0$, as a test function in (13.2.1). Since $\nabla T_1(u_n - T_k(u_n)) = \nabla u_n \chi_{\{k \le |u_n| < k+1\}}$ and $T_1(u_n - T_k(u_n)) = 0$ if $|u_n| \le k$, we can write

$$\int_{B_k} a(x, T_n(u_n), \nabla u_n) \cdot \nabla u_n \le \int_{A_k} |f_n|, \quad \forall\, k \ge s_0$$

where $A_k = \{k \le |u_n|\}$ and $B_k = \{k \le |u_n| < k+1\}$. Assumption (13.1.2) on a implies

$$\gamma k^r \int_{B_k} |\nabla u_n|^2 \le \int_{B_k} a(x, T_n(u_n), \nabla u_n) \cdot \nabla u_n \le \int_{A_k} |f_n|, \quad \forall\, k \ge s_0. \tag{13.2.3}$$

Let $r > 1$; from (13.2.3) we soon obtain the estimate

$$\int_\Omega |\nabla u_n|^2 \le \int_\Omega |\nabla T_{s_0}(u_n)|^2 + \frac{\|f\|_{L^1(\Omega)}}{\gamma} \sum_{k=1}^{\infty} \frac{1}{k^r}.$$

This estimate and point 1 of the above lemma imply that u_n is bounded in $H_0^1(\Omega)$. On the other hand, by Lemma 13.1, $B(u_n)$ belongs to $L^q(\Omega)$ for every $q < \frac{N}{N-2}$. Due to the definition of B there exists a positive constant C such that $|t|^{r+1} \le C(1 + B(t))$. Therefore, u_n belongs to $L^q(\Omega)$ for every $q < \frac{N(r+1)}{N-2}$.

Let $0 < r \le 1$. For every $1 < q < 2$ and $\lambda > 1 - r$, we can write, using Hölder's inequality with exponent $\frac{2}{q}$ and (13.2.3),

$$\begin{aligned}
\int_\Omega |\nabla u_n|^q &= \int_\Omega \frac{|\nabla u_n|^q}{(1+|u_n|)^{\frac{\lambda q}{2}}} (1+|u_n|)^{\frac{\lambda q}{2}} \\
&\le \left[\int_\Omega \frac{|\nabla u_n|^2}{(1+|u_n|)^\lambda} \right]^{\frac{q}{2}} \left[\int_\Omega (1+|u_n|)^{\frac{\lambda q}{2-q}} \right]^{1-\frac{q}{2}} \\
&\le \left[C + \sum_{k=1}^{\infty} \frac{C}{k^{\lambda+r}} \right]^{\frac{q}{2}} \left(\int_\Omega (1+|u_n|)^{\frac{\lambda q}{2-q}} \right)^{1-\frac{q}{2}} \\
&\le C \left[\int_\Omega (1+|u_n|)^{\frac{\lambda q}{2-q}} \right]^{1-\frac{q}{2}}.
\end{aligned}$$

If we set $\lambda = \frac{N(2-q)}{N-q}$, the condition $\lambda > 1 - r$ implies that $q < \frac{N(r+1)}{N-(1-r)}$. The above inequality then yields, by Sobolev's inequality,

$$\left(\int_\Omega |u_n|^{q^*}\right)^{\frac{q}{q^*}} \leq \int_\Omega |\nabla u_n|^q \leq C\left(\int_\Omega (1+|u_n|)^{q^*}\right)^{1-\frac{q}{2}}.$$

Since $\frac{q}{q^*} > 1 - \frac{q}{2}$, we deduce that u_n is bounded in $W_0^{1,q}(\Omega)$ for every $q < \frac{N(r+1)}{N-(1-r)}$. □

Lemma 13.3. *Let $f \in L^m(\Omega)$, $1 < m < \frac{2N}{N+2}$.*
(1) *If $r \geq 1 - \frac{2^*}{m'}$ then the sequence of the solutions u_n to problem* (13.2.1) *is bounded in $H_0^1(\Omega) \cap L^{\frac{Nm(r+1)}{N-2m}}(\Omega)$.*
(2) *If $0 \leq r < 1 - \frac{2^*}{m'}$ then the sequence of the solutions u_n to problem* (13.2.1) *is bounded in $W_0^{1,\frac{Nm(r+1)}{N-m(1-r)}}(\Omega)$.*

Proof. By the same choice of test functions as in Lemma 13.2, we have

$$\gamma k^r \int_{B_k} |\nabla u_n|^2 \leq \int_{B_k} a(x, T_n(u_n), \nabla u_n) \cdot \nabla u_n \leq \int_{A_k} |f_n|, \quad \forall k \geq s_0. \tag{13.2.4}$$

Assume that $r \geq 1 - \frac{2^*}{m'}$. Estimate (13.2.4) and point 1 of Lemma 13.1 give

$$\begin{aligned}\int_\Omega |\nabla u_n|^2 &\leq \int_\Omega |\nabla T_{s_0}(u_n)|^2 + \sum_{k=1}^{\infty} \int_{A_k} \frac{|f_n|}{\gamma k^r} \\ &\leq C + \sum_{k=1}^{\infty} \sum_{h=k}^{\infty} \int_{B_h} \frac{|f_n|}{\gamma k^{1-\frac{2^*}{m'}}} \\ &\leq C + \frac{1}{\gamma} \sum_{h=0}^{\infty} \int_{B_h} |f_n| \sum_{k=0}^{h} \frac{1}{(1+k)^{1-\frac{2^*}{m'}}}.\end{aligned} \tag{13.2.5}$$

Applying in (13.2.5) the inequality $\sum_{k=0}^{h} \frac{1}{(1+k)^s} \leq \frac{(2+h)^{1-s}}{1-s}$, which holds true if $0 < s < 1$, we get

$$\int_\Omega |\nabla u_n|^2 \leq C\left(1 + \int_\Omega |f_n|(2+|u_n|)^{\frac{2^*}{m'}}\right)$$

where C denotes a constant independent of n. By Hölder's and Sobolev's inequalities, it follows that

$$\left(\int_\Omega |u_n|^{2^*}\right)^{\frac{2}{2^*}} \leq \int_\Omega |\nabla u_n|^2 \leq C\left[1 + \left(\int_\Omega (2+|u_n|)^{2^*}\right)^{\frac{1}{m'}}\right].$$

Since $\frac{2}{2^*} > \frac{1}{m'}$, we deduce that u_n is bounded in $L^{2^*}(\Omega)$ and in $H_0^1(\Omega)$.

We now consider as a test function $T_1(|u_n|^r u_n - T_k(|u_n|^r u_n))$ in (13.2.1) in order to have

$$\int_{\{k\le |u_n|^{r+1}<k+1\}} |u_n|^r a(x, T_n(u_n), \nabla u_n)\cdot \nabla u_n \le \int_{\{k\le |u_n|^{r+1}\}} |f_n|\,.$$

Due to (13.1.2), setting $k_1 = s_0^{r+1}$, we have

$$\int_{\{k\le |u_n|^{r+1}<k+1\}} |\nabla(|u_n|^r u_n)|^2 \le C \int_{\{k\le |u_n|^{r+1}\}} |f_n|\,, \qquad \forall\, k\ge k_1\,. \tag{13.2.6}$$

Let $v_n = |u_n|^r u_n$. Then (13.2.6) is equivalent to

$$\int_{\{k\le |u_n|^{r+1}<k+1\}} |\nabla v_n|^2 \le C \int_{\{k\le |v_n|\}} |f_n|\,, \quad \forall\, k\ge k_1\,. \tag{13.2.7}$$

We are going to prove that v_n, that is, $|u_n|^r u_n$ is bounded in $L^{\frac{Nm}{N-2m}}(\Omega)$; thus u_n is bounded in $L^{\frac{Nm(r+1)}{N-2m}}(\Omega)$. By Hölder's inequality with exponent $\frac{2}{q}$ we have

$$\begin{aligned}\int_\Omega |\nabla v_n|^q &= \int_\Omega \frac{|\nabla v_n|^q}{(1+|v_n|)^{\frac{\lambda q}{2}}}(1+|v_n|)^{\frac{\lambda q}{2}}\\ &\le \left[\int_\Omega \frac{|\nabla v_n|^2}{(1+|v_n|)^\lambda}\right]^{\frac{q}{2}}\left[\int_\Omega (1+|v_n|)^{\frac{\lambda q}{2-q}}\right]^{1-\frac{q}{2}}.\end{aligned}$$

Estimate (13.2.7) gives

$$\begin{aligned}\int_\Omega |\nabla v_n|^q &\le C\left[\sum_{n=0}^\infty \frac{1}{(1+n)^\lambda}\sum_{k=n}^\infty \int_{B_k}|f|\right]^{\frac{q}{2}}\left[\int_\Omega (1+|v_n|)^{\frac{\lambda q}{2-q}}\right]^{1-\frac{q}{2}}\\ &\le C\left[\sum_{k=0}^\infty \int_{B_k}|f|\sum_{n=0}^k \frac{1}{(1+n)^\lambda}\right]^{\frac{q}{2}}\left[\int_\Omega (1+|v_n|)^{\frac{\lambda q}{2-q}}\right]^{1-\frac{q}{2}}.\end{aligned}$$

Now, $\sum_{n=0}^k \frac{1}{(1+n)^\lambda}$ can be estimated by $2+k^{1-\lambda}$, which gives

$$\begin{aligned}S^q\left[\int_\Omega |v_n|^{q^*}\right]^{\frac{q}{q^*}} \le \int_\Omega |\nabla v_n|^q &\le C\left[\sum_{k=0}^\infty \int_{B_k}|f|(2+k^{1-\lambda})\right]^{\frac{q}{2}}\left[\int_\Omega (1+|v_n|)^{\frac{\lambda q}{2-q}}\right]^{1-\frac{q}{2}}\\ &\le C\left[\int_\Omega |f| + \int_\Omega |f||v_n|^{1-\lambda}\right]^{\frac{q}{2}}\left[\int_\Omega (1+|v_n|)^{\frac{\lambda q}{2-q}}\right]^{1-\frac{q}{2}}.\end{aligned}$$

By Hölder's inequality with exponent m we get

$$S^q\left[\int_\Omega |v_n|^{q^*}\right]^{\frac{q}{q^*}} \le C\left[\int_\Omega |f| + \|f\|_{L^m(\Omega)}\left(\int_\Omega |v_n|^{(1-\lambda)m'}\right)^{m'}\right]^{\frac{q}{2}}\left[\int_\Omega (1+|v_n|)^{\frac{\lambda q}{2-q}}\right]^{1-\frac{q}{2}}.$$

If $\frac{\lambda q}{2-q} = q^*$, and $q = m^*$, then $q^* = (1-\lambda)m'$. This implies that v_n is bounded in $L^{q^*}(\Omega)$, since $\frac{q}{q^*} \ge 1 - \frac{q}{2} + \frac{q}{2m'}$.

We are left with the case $0 < r < 1 - \frac{2^*}{m'}$. By Hölder's inequality with exponent $\frac{2}{q}$ and (13.2.4), we can write, for every $q < 2$ and $\lambda < 1 - r$,

$$\begin{aligned}\int_\Omega |\nabla u_n|^q &= \int_\Omega \frac{|\nabla u_n|^q}{(1+|u_n|)^{\frac{\lambda q}{2}}}(1+|u_n|)^{\frac{\lambda q}{2}} \\ &\le \left[\int_\Omega \frac{|\nabla u_n|^2}{(1+|u_n|)^\lambda}\right]^{\frac{q}{2}}\left[\int_\Omega (1+|u_n|)^{\frac{\lambda q}{2-q}}\right]^{1-\frac{q}{2}} \\ &\le C\left[\sum_{k=1}^{\infty}\sum_{h=k}^{\infty}\int_{B_h}\frac{|f_n|}{k^{\lambda+r}}\right]^{\frac{q}{2}}\left[\int_\Omega (1+|u_n|)^{\frac{\lambda q}{2-q}}\right]^{1-\frac{q}{2}}.\end{aligned}$$

By exchanging the summation order and using Hölder's inequality with exponent m we have

$$\begin{aligned}\int_\Omega |\nabla u_n|^q &\le C\left[1 + \int_\Omega |f_n|(1+|u_n|)^{1-r-\lambda}\right]^{\frac{q}{2}}\left[\int_\Omega (1+|u_n|)^{\frac{\lambda q}{2-q}}\right]^{1-\frac{q}{2}} \\ &\le C\|f_n\|_{L^m(\Omega)}^{\frac{q}{2}}\left[\int_\Omega (1+|u_n|)^{(1-r-\lambda)m'}\right]^{\frac{q}{2m'}}\left[\int_\Omega (1+|u_n|)^{\frac{\lambda q}{2-q}}\right]^{1-\frac{q}{2}}.\end{aligned}$$

If we set $\lambda = \frac{N(2-q)}{N-q}$ with $q = \frac{Nm(r+1)}{N-m(1-r)}$ we get, by Sobolev's inequality

$$S^q\left[\int_\Omega |u_n|^{q^*}\right]^{\frac{q}{q^*}} \le \int_\Omega |\nabla u_n|^q \le C\left[\int_\Omega (1+|u_n|)^{q^*}\right]^{1-\frac{q}{2m}}.$$

Since $\frac{q}{q^*} \le 1 - \frac{q}{2m}$, this implies that u_n is bounded in $W_0^{1,q}(\Omega)$ with $q = \frac{Nm(r+1)}{N-m(1-r)}$.

□

Lemma 13.4. *Let* $f \in L^m(\Omega)$, $m \geq \frac{2N}{N+2}$.

(1) *If* $m > \frac{N}{2}$, *then the sequence of the solutions* u_n *to* (13.2.1) *is bounded in* $H_0^1(\Omega) \cap L^\infty(\Omega)$.
(2) *If* $\frac{2N}{N+2} \leq m < \frac{N}{2}$, *then the sequence of the solutions* u_n *to* (13.2.1) *is bounded in* $H_0^1(\Omega) \cap L^{\frac{Nm(r+1)}{N-2m}}(\Omega)$.

Proof. In the first case, one can choose as a test function $G_k(u_n)$. One has by the ellipticity of a

$$\alpha \int_\Omega |\nabla G_k(u_n)|^2 \leq \int_\Omega f G_k(u_n) .$$

By Remark 6.8 this estimate gives a uniform bound for the $L^\infty(\Omega)$ norm of u_n.

In the second case, we use the same technique as in Theorem 6.9. Consider as a test function

$$\frac{|T_k(u_n)|^{2\lambda} T_k(u_n)}{2\lambda + 1},$$

with $\lambda > 0$. We will study separately the two sides of the weak formulation. For the left-hand side we have, for $|T_k(u_n)| \geq s_0$,

$$\frac{1}{2\lambda+1} \int_\Omega a(x, T_n(u_n), \nabla u_n) \cdot \nabla(|T_k(u_n)|^{2\lambda} T_k(u_n)) \geq \alpha S^2 \left(\int_\Omega |T_k(u_n)|^{(\lambda + \frac{r}{2} + 1)2^*} \right)^{\frac{2}{2^*}}$$

by assumption (13.1.3) on a and Sobolev's inequality. For the right-hand side, we get

$$\frac{1}{2\lambda+1} \int_\Omega f |T_k(u_n)|^{2\lambda} T_k(u_n) \leq \frac{1}{2\lambda+1} \left(\int_\Omega |T_k(u_n)|^{(2\lambda+1)m'} \right)^{\frac{1}{m'}} \|f\|_{L^m(\Omega)}$$

by using Hölder's inequality with exponent m. We have thus proved that

$$\alpha S^2 (2\lambda + 1) \left[\int_\Omega |T_k(u_n)|^{(\lambda + \frac{r}{2} + 1)2^*} \right]^{\frac{2}{2^*}} \leq \left[\int_\Omega |T_k(u_n)|^{(2\lambda+1)m'} \right]^{\frac{1}{m'}} \|f\|_{L^m(\Omega)} .$$

Now it is sufficient to choose λ such that $(\lambda + \frac{r}{2} + 1)2^* = m'(2\lambda + 1)$, that is,

$$\lambda = \frac{-m(N + 2 + rN) + N(r + 2)}{4m - 2N}$$

Using that $\frac{2}{2^*} - \frac{1}{m'} > 0$, one has

$$\|T_k(u_n)\|_{L^{m^{**}(r+1)}(\Omega)} \leq C \|f\|_{L^m(\Omega)}.$$

Fatou's Lemma, as $k \to \infty$, implies the result. □

Theorem 13.5. *Let $f \in L^1(\Omega)$. There exists a function u such that $T_k(u) \in H_0^1(\Omega)$, $a(x,u,\nabla u)$ belongs to $L^q(\Omega)$ for every $q < \frac{N}{N-1}$ and u solves (13.1.1) in the sense of distributions.*
(1) *If $r > 1$, then u belongs to $H_0^1(\Omega) \cap L^s(\Omega)$, $s < \frac{N(r+1)}{N-2}$.*
(2) *If $0 \le r \le 1$, then u belongs to $W_0^{1,q}(\Omega)$, $q < \frac{N(r+1)}{N-1+r}$.*

Let $f \in L^m(\Omega)$, $1 < m < \frac{2N}{N+2}$.
(1) *If $r \ge 1 - \frac{2^*}{m'}$ then u belongs to $H_0^1(\Omega) \cap L^{\frac{Nm(r+1)}{N-2m}}(\Omega)$.*
(2) *If $0 \le r < 1 - \frac{2^*}{m'}$ then u belongs to $W_0^{1,\frac{Nm(r+1)}{N-m(1-r)}}(\Omega)$.*

Let $f \in L^m(\Omega)$, $m \ge \frac{2N}{N+2}$.
(1) *If $m > \frac{N}{2}$, then u belongs to $H_0^1(\Omega) \cap L^\infty(\Omega)$.*
(2) *If $\frac{2N}{N+2} \le m < \frac{N}{2}$, then u belongs to $H_0^1(\Omega) \cap L^{\frac{Nm(r+1)}{N-2m}}(\Omega)$.*

Proof. By Lemmata 13.1 and 11.27, there exists a function u such that $T_k(u)$ belongs to $H_0^1(\Omega)$ and, up to subsequences, u_n converges to u a.e. in Ω, $T_k(u_n)$ converges to $T_k(u)$ weakly in $H_0^1(\Omega)$ and strongly in $L^2(\Omega)$ and a.e. in Ω. Again by Lemma 13.1, always reasoning up to subsequences, $a(x, T_n(u_n), \nabla u_n)$ is weakly convergent in $L^q(\Omega)$, $q < \frac{N}{N-1}$.

We are going to prove that ∇u_n converges to ∇u a.e. in Ω. This will allow us to pass to the limit in (13.2.1) as n goes to infinity with test functions in $C_0^\infty(\Omega)$ in order to obtain a distributional solution to problem (13.1.1). We will use a technique similar to that one used in Lemma 11.3. For $\varepsilon > 0$, $n \ge k > 0$, we take $T_\varepsilon(u_n - T_k(u))$ as a test function in (13.2.1); we have

$$\int_\Omega a(x, T_n(u_n), \nabla u_n) \cdot \nabla T_\varepsilon(u_n - T_k(u)) \le \varepsilon \|f\|_{L^1(\Omega)}. \tag{13.2.8}$$

Since $\nabla T_\varepsilon(u_n - T_k(u)) = 0$ if $|u_n| > k + \varepsilon$ and $a(x,s,\xi) \cdot \xi \ge 0$, we have

$$\begin{aligned}\int_\Omega & a(x, T_n(u_n), \nabla u_n) \cdot \nabla T_\varepsilon(u_n - T_k(u)) \\ &\ge \int_\Omega a(x, T_k(u_n), \nabla T_k(u_n)) \cdot \nabla T_\varepsilon(T_k(u_n) - T_k(u)) \\ &\quad - \int_{\{|u_n|>k\}} |a(x, T_{k+\varepsilon}(u_n), \nabla T_{k+\varepsilon}(u_n))| \, |\nabla T_k(u)|.\end{aligned}$$

Now, $|\nabla T_k(u)|\chi_{\{|u_n|>k\}}$ strongly converges to 0 in $L^2(\Omega)$, so that last term in the above inequality goes to 0 as $n \to \infty$ for every fixed $\varepsilon > 0$. We will denote by δ_n^ε any quantity converging to 0 as $n \to \infty$, for every fixed $\varepsilon > 0$. Thus by (13.2.8) it follows that

$$\int_\Omega a(x, T_k(u_n), \nabla T_k(u_n)) \cdot \nabla T_\varepsilon(T_k(u_n) - T_k(u)) \le \varepsilon \|f\|_{L^1(\Omega)} + \delta_n^\varepsilon. \tag{13.2.9}$$

Now, let $0 < \theta < 1$ and set $E_k^\varepsilon = \{\, |T_k(u_n) - T_k(u)| > \varepsilon\}$. We have

$$\int_\Omega \{(a(x, T_k(u_n), \nabla T_k(u_n)) - a(x, T_k(u_n), \nabla T_k(u))) \cdot \nabla(T_k(u_n) - T_k(u))\}^\theta$$
$$= \int_\Omega \{(a(x, T_k(u_n), \nabla T_k(u_n)) - a(x, T_k(u_n), \nabla T_k(u))) \cdot \nabla T_\varepsilon(T_k(u_n) - T_k(u))\}^\theta$$
$$+ \int_{E_k^\varepsilon} \{(a(x, T_k(u_n), \nabla T_k(u_n)) - a(x, T_k(u_n), \nabla T_k(u))) \cdot \nabla(T_k(u_n) - T_k(u))\}^\theta .$$

By Hölder's inequality with exponent $\frac{1}{\theta}$ in the second integral we get, from (13.2.9),

$$\begin{aligned} &\int_\Omega \{(a(x, T_k(u_n), \nabla T_k(u_n)) - a(x, T_k(u_n), \nabla T_k(u))) \cdot \nabla(T_k(u_n) - T_k(u))\}^\theta \\ &\quad \le \operatorname{meas}(\Omega)^{1-\theta} \left(\varepsilon \|f\|_{L^1(\Omega)} + \delta_n^\varepsilon - \int_\Omega a(x, T_k(u_n), \nabla T_k(u)) \cdot \nabla T_\varepsilon(T_k(u_n) - T_k(u)) \right)^\theta \\ &\quad + \int_{E_k^\varepsilon} \{(a(x, T_k(u_n), \nabla T_k(u_n)) - a(x, T_k(u_n), \nabla T_k(u))) \cdot \nabla(T_k(u_n) - T_k(u))\}^\theta . \end{aligned} \tag{13.2.10}$$

We are going to study the right-hand side of (13.2.10). Since $T_k(u_n)$ is bounded in $H_0^1(\Omega)$ and weakly converges to $T_k(u)$, by the assumptions on a the first integral goes to zero as n goes to infinity. Moreover the sequence $\{|a(x, T_k(u_n), \nabla T_k(u_n)) - a(x, T_k(u_n), \nabla T_k(u))|\}$ is bounded in $L^2(\Omega)$ for every fixed $k > 0$. Thus, using Hölder's inequality on the last integral, we get

$$\begin{aligned} &\int_\Omega \{(a(x, T_k(u_n), \nabla T_k(u_n)) - a(x, T_k(u_n), \nabla T_k(u))) \nabla(T_k(u_n) - T_k(u))\}^\theta \\ &\quad \le \operatorname{meas}(\Omega)^{1-\theta} \left(\varepsilon \|f\|_{L^1(\Omega)} + 2\delta_n^\varepsilon\right)^\theta + C\operatorname{meas}(\{|T_k(u_n) - T_k(u)| > \varepsilon\})^{1-\theta} . \end{aligned}$$

Since $T_k(u_n)$ converges to $T_k(u)$ in measure, letting first n go to infinity and then letting ε go to zero, we find

$$\begin{aligned} \lim_{n\to+\infty} \int_\Omega \{(a(x, T_k(u_n), \nabla T_k(u_n)) \\ - a(x, T_k(u_n), \nabla T_k(u))) \cdot \nabla(T_k(u_n) - T_k(u))\}^\theta = 0 . \end{aligned}$$

We deduce that $\nabla T_k(u_n)$ converges a.e. to $\nabla T_k(u)$ for every $k > 0$. This implies that ∇u_n converges to ∇u a.e. in Ω, so that $a(x, T_n(u_n), \nabla u_n)$ converges to $a(x, u, \nabla u)$ strongly in $L^q(\Omega)$ for every $q < \frac{N}{N-1}$.

The above argument implies the existence of a distributional solution u to problem (13.1.1). The regularity of u follows from the a.e. convergence of u_n to u and Lemmata 13.2, 13.3, and 13.4. $\square$

14 A problem with degenerate coercivity

14.1 Introduction

In this chapter we will treat some problems defined by an elliptic operator with degenerate coercivity, that is, an elliptic operator which does not satisfy assumption 2 of Leray–Lions theorem (Theorem 5.1). More precisely let Ω be a bounded, open subset of $\mathbb{R}^N$, with $N > 2$. We assume that $a(x,s) : \Omega \times \mathbb{R} \to \mathbb{R}$ is a Carathéodory function satisfying the following conditions:

$$\frac{\alpha}{(1+|s|)^{\theta}} \le a(x,s) \le \beta, \tag{14.1.1}$$

for almost every $x \in \Omega$, for every $s \in \mathbb{R}$, where α, β, and θ are positive constants. Note that, because of assumption (14.1.1), the differential operator $A(v) = -\operatorname{div}(a(x,v)\nabla v)$ is not coercive on $H_0^1(\Omega)$, even if it is well defined between $H_0^1(\Omega)$ and its dual. To see that it is sufficient to consider the sequence $u_n(x) = |x|^{\frac{n(2-N)}{2(n+1)}} - 1$ defined in $\Omega = B_1(0)$. It satisfies $\|u_n\|_{H_0^1(\Omega)} \to \infty$ and, at the same time,

$$\frac{1}{||u_n||_{H_0^1(\Omega)}} \int_\Omega \frac{|\nabla u_n|^2}{(1+|u_n|)^p} \to 0.$$

We will begin by studying the existence and regularity of solutions to

$$\begin{cases} -\operatorname{div}(a(x,u)\nabla u) = f, & \text{in } \Omega, \\ u = 0, & \text{on } \partial\Omega, \end{cases} \tag{14.1.2}$$

where the source f belongs to $L^m(\Omega)$, with $m \ge 1$.

In the case where $\theta < 1$ we will prove the following results (following [10]):

(1) if $m > \frac{N}{2}$ there exists a solution u in $H_0^1(\Omega) \cap L^\infty(\Omega)$;
(2) if $\frac{2N}{N+2-\theta(N-2)} \le m < \frac{N}{2}$ there exists a solution u in $H_0^1(\Omega) \cap L^{m^{**}(1-\theta)}(\Omega)$;
(3) if $\frac{N}{N+1-\theta(N-1)} < m < \frac{2N}{N+2-\theta(N-2)}$ there exists a solution u in $W_0^{1,q}(\Omega)$, $q = \frac{Nm(1-\theta)}{N-m(1+\theta)} < 2$.

If we weaken the summability hypotheses on f, then the gradient of u (and even u itself) may no longer be in $L^1(\Omega)$. However, it is possible to give a meaning to solution for problem (14.1.2), using the concept of entropy solutions (see [10] for further details and references).

In the case where $\theta > 1$, some nonexistence phenomena appear. In Section 14.3 we will prove that for constant sources sufficiently large, there is no solution. However, we will prove an existence theorem for "small" sources in Theorem 14.10.

The existence of solutions to problem (14.1.2) can be recovered for every value of θ by adding a lower order term. More precisely we will study the existence and regularity of solutions to

$$\begin{cases} -\mathrm{div}(a(x,u)\nabla u) + u = f, & \text{in } \Omega, \\ u = 0, & \text{on } \partial\Omega, \end{cases} \tag{14.1.3}$$

proving the following results (see [8] and [19]):
(1) if $m \geq \theta + 2$, $\theta > 0$, there exists a solution u in $H_0^1(\Omega) \cap L^m(\Omega)$;
(2) if $\frac{\theta+2}{2} < m < \theta + 2$, $\theta > 0$, there exists a solution u in $W_0^{1,\frac{2m}{\theta+2}}(\Omega) \cap L^m(\Omega)$;
(3) if $m > \theta\frac{N}{2}$ and $\theta > 1$, there exists a solution u in $H_0^1(\Omega) \cap L^\infty(\Omega)$.

In the case where $\theta = 2 = m$, we will prove the existence of a unique $W_0^{1,1}(\Omega)$ solution obtained by approximation and nonexistence in the case of a Dirac mass source.

14.2 The case $0 < \theta < 1$

In this section, we are going to study problem (14.1.2) under the hypothesis that $\theta < 1$. We will work by approximation. Let $f_n = T_n(f)$. We define the following sequence of problems:

$$\begin{cases} -\mathrm{div}(a(x,T_n(u_n))\nabla u_n) = f_n, & \text{in } \Omega, \\ u_n = 0, & \text{on } \partial\Omega. \end{cases} \tag{14.2.1}$$

We remark that, due to the fact that we have truncated the second variable of a, problems (14.2.1) are coercive, so that the existence of weak solutions u_n in $H_0^1(\Omega) \cap L^\infty(\Omega)$ to (14.2.1) is assured by Theorems 5.1 and 6.6.

Lemma 14.1. *Let $\theta > 0$ and let f belongs to $L^m(\Omega)$, $m > \frac{N}{2}$. Assume that u_n is a solution to problem (14.2.1) with $f_n = f$ for every $n \in \mathbb{N}$. Then the norms of u_n in $L^\infty(\Omega)$ and in $H_0^1(\Omega)$ are bounded.*

Proof. We define, for s in $\mathbb{R}$ and for $k > 0$

$$H(s) = \int_0^s \frac{1}{(1+|t|)^\theta}\, dt.$$

For $k > 0$, if we take $G_k(H(u_n))$ as a test function in (14.2.1) and use assumption (14.1.1), we obtain

$$\alpha \int_{A_k} |\nabla(H(u_n))|^2 \leq \int_{A_k} f\, G_k(H(u_n)), \tag{14.2.2}$$

where we have set

$$A_k = \{x \in \Omega : |H(u_n(x))| > k\}.$$

By Remark (6.8), there exists a constant $C > 0$ (depending on θ, m, N, α, meas(Ω) and on the norm of f in $L^m(\Omega)$) such that

$$\|H(u_n)\|_{L^\infty(\Omega)} \le C\,.$$

The limits $\lim_{s\to+\infty} H(s) = +\infty$, $\lim_{s\to-\infty} H(s) = -\infty$ yield a bound for u_n in $L^\infty(\Omega)$:

$$\|u_n\|_{L^\infty(\Omega)} \le C\,. \tag{14.2.3}$$

The uniform estimate of u_n in $H_0^1(\Omega)$ is now very easy. Taking u_n as test function in (14.2.1), one obtains, by assumption (14.1.1) and (14.2.3)

$$\frac{\alpha}{(1+C)^\theta}\int_\Omega |\nabla u_n|^2 \le \int_\Omega f\,u_n \le C\|f\|_{L^1(\Omega)}\,.$$

□

Theorem 14.2. *Let f be a function in $L^m(\Omega)$, with $m > \frac{N}{2}$. Then there exists a weak solution $u \in H_0^1(\Omega)\cap L^\infty(\Omega)$ to problem (14.1.2), that is,*

$$\int_\Omega a(x,u)\nabla u\cdot\nabla\varphi = \int_\Omega f\varphi\,, \quad \forall\,\varphi\in H_0^1(\Omega)\,.$$

Proof. In the above lemma we proved the existence of a positive constant C such that $\|u_n\|_{L^\infty(\Omega)} \le C$ for every $n\in\mathbb{N}$. It is then sufficient to take $\nu\in\mathbb{N}$ with $\nu > C$, so that $T_\nu(u_\nu) = u_\nu$. Then u_ν is a weak solution to problem (14.1.2) belonging to $H_0^1(\Omega)\cap L^\infty(\Omega)$. □

Remark 14.3. We observe that in Theorem 14.2 we do not assume that $\theta < 1$. This will be useful to prove Theorem 14.10.

We now weaken the summability of the source f.

Lemma 14.4. *Let $0 < \theta < 1$ and $\frac{2N}{N+2-\theta(N-2)} \le m < \frac{N}{2}$. Let u_n be the solutions to (14.2.1). Then the norms of u_n in $L^{m^{**}(1-\theta)}(\Omega)$ and in $H_0^1(\Omega)$ are bounded.*

Proof. We consider $[(1+|u_n|)^p - 1]\operatorname{sgn}(u_n)$, with $p = \frac{(1-\theta)N(m-1)}{N-2m}$, as a test function in (14.2.1). We observe that p satisfies $pm' = (p+1-\theta)2^*/2 = m^{**}(1-\theta) > 1$. By assumption (14.1.1), we have

$$\begin{aligned}\frac{4\alpha}{(p+1-\theta)^2}\int_\Omega |\nabla((1+|u_n|)^{\frac{p+1-\theta}{2}}-1)|^2 &= \alpha\int_\Omega \frac{|\nabla u_n|^2}{(1+|u_n|)^{\theta-p+1}}\\ &\le C\int_\Omega |f||u_n|^p\,.\end{aligned} \tag{14.2.4}$$

Hölder's inequality on the right-hand side and Sobolev's one on the left-hand side imply

$$\left[\int_\Omega |u_n|^{(p+1-\theta)2^*/2}\right]^{\frac{2}{2^*}} \le C\int_\Omega \frac{|\nabla u_n|^2}{(1+|u_n|)^{\theta-p+1}} \le C\|f\|_{L^m(\Omega)}\left[\int_\Omega |u_n|^{pm'}\right]^{\frac{1}{m'}}\,.$$

Since $\frac{2}{2^*} > \frac{1}{m'}$, the above estimate gives that the sequence u_n is bounded in $L^{m^{**}(1-\theta)}(\Omega)$. From this uniform estimate we deduce that the sequence $\int_\Omega |\nabla u_n|^2(1+|u_n|)^{p-1-\theta}$ is bounded. As $p-1-\theta > 0$, by the assumptions on m, this implies a uniform bound for ∇u_n in $L^2(\Omega)$. □

Theorem 14.5. *Let f be a function in $L^m(\Omega)$, with $\frac{2N}{N+2-\theta(N-2)} < m < \frac{N}{2}$. Then there exists a weak solution $u \in H_0^1(\Omega) \cap L^{m^{**}(1-\theta)}(\Omega)$ to problem (14.1.2), that is,*

$$\int_\Omega a(x,u)\nabla u \cdot \nabla\varphi = \int_\Omega f\varphi, \quad \forall\, \varphi \in H_0^1(\Omega).$$

Proof. The estimate for u_n in $H_0^1(\Omega)$ implies that, up to a subsequence, u_n converges to some function $u \in H_0^1(\Omega)$ weakly in $H_0^1(\Omega)$ and a.e. in Ω. By Lebesgue's theorem the coefficient $a(x,u_n)$ converges to $a(x,u)$ in $L^q(\Omega)$, for any q, due to assumption (14.1.1). Thus it is possible to pass to the limit in (14.2.1) in order to obtain the existence of a weak solution u of problem (14.1.2). □

The last result gives the existence of distributional solutions in the case where $f \in L^m(\Omega)$, with $\frac{N}{N+1-\theta(N-1)} < m < \frac{2N}{N(1-\theta)+2(\theta+1)}$.

Lemma 14.6. *Assume that $f \in L^m(\Omega)$, with $\frac{N}{N+1-\theta(N-1)} < m < \frac{2N}{N(1-\theta)+2(\theta+1)}$. Then the sequence of the solutions u_n to (14.2.1) is bounded in $W_0^{1,q}(\Omega)$, with $q = \frac{Nm(1-\theta)}{N-m(1+\theta)}$.*

Proof. Choosing $T_1(G_k(u_n))$ as a test function in (14.2.1), we get, by assumption (14.1.1)

$$\alpha \int_{B_k} |\nabla u_n|^2 \le (2+k)^\theta \int_{A_k} |f|, \tag{14.2.5}$$

where $A_k = \{|u_n| \ge k\}$ and $B_k = \{k \le |u_n| < k+1\}$. Let $\lambda = \frac{2N-2m(1+\theta)-Nm(1-\theta)}{N-2m}$. Then, by (14.2.5), we have

$$\begin{aligned}\int_\Omega \frac{|\nabla u_n|^2}{(1+|u_n|)^\lambda} &= \sum_{k=0}^{+\infty} \int_{B_k} \frac{|\nabla u_n|^2}{(1+|u_n|)^\lambda} \\ &\le \sum_{k=0}^{+\infty} \frac{1}{(1+k)^\lambda} \int_{B_k} |\nabla u_n|^2 \\ &\le \frac{1}{\alpha} \sum_{k=0}^{+\infty} (2+k)^\theta (1+k)^{-\lambda} \int_{A_k} |f| \\ &\le \frac{1}{\alpha} \sum_{k=0}^{+\infty} (2+k)^{\theta-\lambda} \sum_{h=k}^{+\infty} \int_{B_k} |f|.\end{aligned} \tag{14.2.6}$$

By changing the order of summation we obtain

$$\begin{aligned}\int_\Omega \frac{|\nabla u_n|^2}{(1+|u_n|)^\lambda} &\le \frac{1}{\alpha}\sum_{h=0}^{+\infty}\int_{B_h}|f|\sum_{k=0}^{h}(2+k)^{\theta-\lambda}\\ &\le C\sum_{h=0}^{+\infty}(3+h)^{1+\theta-\lambda}\int_{B_h}|f|\\ &\le C\sum_{h=0}^{+\infty}\int_{B_h}|f|(3+|u_n|)^{1+\theta-\lambda}\\ &= C\int_\Omega |f|(3+|u_n|)^{1+\theta-\lambda}\end{aligned}$$

where we have used that $\sum_{k=0}^{h}(2+k)^s \le C\int_0^{h+1}(2+t)^s dt \le C(3+h)^{s+1}$. By Hölder's inequality we deduce that

$$\begin{aligned}\int_\Omega \frac{|\nabla u_n|^2}{(1+|u_n|)^\lambda} &\le C\|f\|_{L^m(\Omega)}\left[\int_\Omega (3+|u_n|)^{(1+\theta-\lambda)m'}\right]^{\frac{1}{m'}}\\ &\le C + C\left[\int_\Omega |u_n|^{(1+\theta-\lambda)m'}\right]^{\frac{1}{m'}},\end{aligned}\tag{14.2.7}$$

since $1+\theta-\lambda>0$. By the Sobolev inequality and the Hölder one with exponent $\frac{2}{q}$, we have

$$\begin{aligned}S^q\left[\int_\Omega |u_n|^{q^*}\right]^{\frac{q}{q^*}} &\le \int_\Omega |\nabla u_n|^q\\ &= \int_\Omega \frac{|\nabla u_n|^q}{(1+|u_n|)^{\frac{\lambda q}{2}}}(1+|u_n|)^{\frac{\lambda q}{2}}\\ &\le \left[\int_\Omega \frac{|\nabla u_n|^2}{(1+|u_n|)^\lambda}\right]^{\frac{q}{2}}\left[\int_\Omega (1+|u_n|)^{\frac{\lambda q}{2-q}}\right]^{\frac{2-q}{2}}.\end{aligned}\tag{14.2.8}$$

Estimate (14.2.7) implies

$$\left[\int_\Omega |u_n|^{q^*}\right]^{\frac{q}{q^*}} \le \left\{C + C\left[\int_\Omega |u_n|^{(1+\theta-\lambda)m'}\right]^{\frac{q}{2m'}}\right\}\left\{1+\left[\int_\Omega |u_n|^{\frac{\lambda q}{2-q}}\right]^{\frac{2-q}{2}}\right\}.$$

Since $\frac{\lambda q}{2-q} = q^*$ and $(1+\theta-\lambda)m' \le q^*$, one obtains that the sequence u_n is bounded in $L^{q^*}(\Omega)$. From (14.2.8) and (14.2.7) we deduce that

$$\int_\Omega |\nabla u_n|^q \le \left\{ C + C\left[\int_\Omega |u_n|^{(1+\theta-\lambda)m'}\right]^{\frac{q}{2m'}} \right\} \left\{ 1 + \left[\int_\Omega |u_n|^{\frac{\lambda q}{2-q}}\right]^{\frac{2-q}{2}} \right\},$$

that is, the sequence u_n is bounded in $W_0^{1,q}(\Omega)$. □

Passing to the limit in problems (14.2.1), as in Theorem 14.5, it is easy to prove the following result. Observe that the choice of test functions is different from the above theorems, due to the regularity of the solutions.

Theorem 14.7. *Assume that $f \in L^m(\Omega)$, with $\frac{N}{N+1-\theta(N-1)} < m \le \frac{2N}{N(1-\theta)+2(\theta+1)}$. Let $q = \frac{Nm(1-\theta)}{N-m(1+\theta)}$. Then there exists a distributional solution $u \in W_0^{1,q}(\Omega)$ to problem* (14.1.2)*, that is,*

$$\int_\Omega a(x,u)\nabla u \cdot \nabla\varphi = \int_\Omega f\varphi, \quad \forall\, \varphi \in C_0^\infty(\Omega).$$

Remark 14.8. We observe that $q^* = \frac{Nm(1-\theta)}{N-2m}$, which is the summability found in Theorem 14.5.

We end this section by studying under which conditions on the summability of the source f the solution u can be chosen as a test function in problem (14.1.2).

Proposition 14.9. *Assume that $m \ge \frac{N(2-\theta)}{N+2-N\theta}$. Let u be a solution to problem* (14.1.2) *found by approximation. Then*

$$\int_\Omega a(x,u)\,|\nabla u|^2 = \int_\Omega f u.$$

Proof. Step I: We are going to prove that

$$\int_\Omega a(x,u)\,|\nabla u|^2 \le \int_\Omega f u < \infty. \tag{14.2.9}$$

Let u_n be a solution to (14.2.1). We have proved in Lemmata 14.1, 14.4 and 14.6, that the sequence u_n is bounded in $L^r(\Omega)$, where $r = \frac{Nm(1-\theta)}{N-2m} = q^*$, with $q = \frac{Nm(1-\theta)}{N-m(1+\theta)}$. Moreover, choosing $T_k(u_n)$ as a test function in (14.2.1), we obtain

$$\int_\Omega a(x, T_n(u_n))\,|\nabla T_k(u_n)|^2 = \int_\Omega f_n\, T_k(u_n). \tag{14.2.10}$$

By using (14.1.1) on the left-hand side and Hölder's inequality on the right one we have

$$\alpha \int_{A_k} \frac{|\nabla T_k(u_n)|^2}{(1+k)^\theta} \le k\|f\|_{L^m(\Omega)},$$

that is, for a fixed $k > 0$, the sequence $T_k(u_n)$ is bounded in $H_0^1(\Omega)$. It is easy to see that

$$2\int_\Omega a(x,T_n(u_n))\,\nabla T_k(u_n)\cdot\nabla T_k(u) - \int_\Omega a(x,T_n(u_n))\,|\nabla T_k(u)|^2 \\ \le \int_\Omega a(x,T_n(u_n))\,|\nabla T_k(u_n)|^2 .$$

Since a is bounded by assumption (14.1.1), the left-hand side of the above inequality converges to $\int_\Omega a(x,u)\,|\nabla T_k(u)|^2$. Therefore, by (14.2.10), we have

$$\int_\Omega a(x,u)\,|\nabla T_k(u)|^2 \le \int_\Omega f\,T_k(u) .$$

At the limit as $k \to \infty$, the right-hand side tends to $\int_\Omega fu$, which is finite by Hölder's inequality, since $q^* = r \ge m'$. By Fatou's Lemma on the left-hand side we obtain (14.2.9).

Step II: Let φ_n be a sequence of $C_0^\infty(\Omega)$ functions, converging to $T_k(u)$ in $H_0^1(\Omega)$ and $*$-weakly in $L^\infty(\Omega)$. Then

$$\int_\Omega a(x,u)\,\nabla u\cdot\nabla\varphi_n = \int_\Omega f\,\varphi_n .$$

The sequence $a(x,u)\,\nabla u\cdot\nabla\varphi_n$ converges to $a(x,u)|\nabla T_k(u)|^2$ a.e. in Ω. Moreover, if E is a measurable subset of Ω, we have, by Hölder's inequality and (14.1.1)

$$\int_E a(x,u)\,\nabla u\cdot\nabla\varphi_n \le \left(\int_E a(x,u)\,|\nabla u|^2\right)^{\frac12}\left(\int_E \beta\,|\nabla\varphi_n|^2\right)^{\frac12} .$$

By Vitali's theorem (Theorem 3.2) the right-hand side converges to 0, as $\operatorname{meas}(E) \to 0$, due to (14.2.9) and the hypotheses on φ_n. At the limit as $n \to \infty$, Vitali's theorem implies that

$$\int_\Omega a(x,u)\,|\nabla T_k(u)|^2 = \int_\Omega f\,T_k(u) .$$

It is now sufficient to pass to the limit as $k \to \infty$. □

14.3 The case $\theta > 1$: existence and nonexistence

If $\theta > 1$, some nonexistence phenomena occur for problem (14.1.2), even for constant sources. We observe that one would expect these kinds of sources to give bounded solutions, if one thinks to Theorem 6.6.

Assume that u is a solution to

$$\begin{cases} -\operatorname{div}\left(\dfrac{\nabla u}{(1+|u|)^{\theta}}\right) = A > 0\,, & \text{in} \quad B_1(0)\,, \\ u = 0\,, & \text{on} \quad \partial B_1(0)\,. \end{cases}$$

Then

$$v(x) = \frac{1-(1+|u|)^{1-\theta}}{\theta - 1}$$

solves $-\Delta v = A$. Observe that

$$v \le \frac{1}{\theta - 1} \quad \text{in} \quad B_1(0)\,. \tag{14.3.1}$$

Now, the $H_0^1(\Omega)$ solution to $-\Delta v = A$ is $v = \frac{A}{2N}(1-|x|^2)$. If A is sufficiently large, there exists a subset E_0 of $B_1(0)$, with $\operatorname{meas}(E_0) > 0$, where $v > \frac{1}{\theta-1}$. This contradicts (14.3.1).

However, if A is sufficiently small a solution exists, as stated in the following existence result in the case where $\theta > 1$.

Theorem 14.10. *Let Ω be a bounded open subset of $\mathbb{R}^N$ with $\partial\Omega$ of class C^1. Assume that a satisfies* (14.1.1) *with $\theta > 1$. Let f belong to $L^m(\Omega)$, with $m > \frac{N}{2}$. Then there exists $\lambda^* > 0$ such that, if $|\lambda| < \lambda^*$, problem*

$$\begin{cases} -\operatorname{div}(a(x,u)\nabla u) = \lambda f\,, & \text{in} \quad \Omega\,, \\ u = 0\,, & \text{on} \quad \partial\Omega\,, \end{cases} \tag{14.3.2}$$

has a bounded $H_0^1(\Omega)$ solution.

Proof. We will use Theorem 2.10 (Schauder's fixed point Theorem) to prove the result. We define $S : L^\infty(\Omega) \to L^\infty(\Omega)$ as the operator which maps v into z, where $z \in H_0^1(\Omega)$ is the weak solution to

$$-\operatorname{div}(a(x,v)\nabla z) = \lambda f\,. \tag{14.3.3}$$

S is well defined, due to Theorem 14.2 and Remark 14.3.

Step I: We prove that S has an invariant convex, closed, bounded set. Let us consider $G_k(z)$ as a test function in (14.3.3). By assumption (14.1.1), one has

$$\alpha \int_\Omega \frac{|\nabla G_k(z)|^2}{(1+|v|)^\theta} \le \int_\Omega a(x,v)|\nabla G_k(z)|^2 \le \lambda \int_\Omega f\, G_k(z)\,.$$

Since v is bounded, we deduce that

$$\alpha \int_\Omega |\nabla G_k(z)|^2 \le \lambda\,(1+\|v\|_{L^\infty(\Omega)})^\theta \int_\Omega f\, G_k(z)\,.$$

By Remark (6.8), there exists a positive constant C such that

$$\|z\|_{L^\infty(\Omega)} \leq C\,|\lambda|\,\|f\|_{L^m(\Omega)}(1+\|v\|_{L^\infty(\Omega)})^\theta .$$

We define $h:\mathbb{R}^+ \to \mathbb{R}$ by

$$h(R) = R - C\,\|f\|_{L^m(\Omega)}\,|\lambda|(1+R)^\theta .$$

We will prove that there exists R_0 such that $h(R_0) > 0$. The maximum of h is attained at

$$\frac{1}{(C|\lambda|\theta\|f\|_{L^m(\Omega)})^{\frac{1}{\theta-1}}} - 1 = R_1 ;$$

this value is well defined if

$$\frac{1}{(C|\lambda|\theta\|f\|_{L^m(\Omega)})^{\frac{1}{\theta-1}}} > 1 ,$$

that is,

$$|\lambda| \leq \frac{1}{C\theta\|f\|_{L^m(\Omega)}} .$$

Now,

$$h(R_1) = \frac{1}{(|\lambda|\theta\|f\|_{L^m(\Omega)})^{\frac{1}{\theta-1}}} - 1 - C\lambda\left(\frac{1}{(C|\lambda|\theta\|f\|_{L^m(\Omega)})^{\frac{\theta}{\theta-1}}}\right) > 0$$

if and only if

$$\frac{\theta-1}{\theta^{\frac{\theta}{\theta-1}}} > (C|\lambda|\|f\|_{L^m(\Omega)})^{\frac{1}{\theta-1}} .$$

Therefore it is sufficient to choose λ such that

$$|\lambda| < \min\left\{\frac{(\theta-1)^{\theta-1}}{\theta^\theta\,C\|f\|_{L^m(\Omega)}}, \frac{1}{C\,\|f\|_{L^m(\Omega)}\,\theta}\right\} = \frac{(\theta-1)^{\theta-1}}{\theta^\theta\,\|f\|_{L^m(\Omega)}C} .$$

In this way the closed $L^\infty(\Omega)$ ball B of radius R_1 centered in 0 is invariant under S. It is clear that B is bounded and convex.

Step II: We will show that S is completely continuous. We are going to prove that if $v_n \to v$ in $L^\infty(\Omega)$, then $S(v_n) = z_n \to S(v) = z$ in $L^\infty(\Omega)$. By subtracting side by side of the equalities

$$\int_\Omega a(x,v_n)\nabla z_n \cdot \nabla w = \int_\Omega \lambda f w$$

and

$$\int_\Omega a(x,v)\nabla z \cdot \nabla w = \int_\Omega \lambda f w ,$$

we obtain

$$\int_\Omega a(x,v_n)\nabla z_n \cdot \nabla w = \int_\Omega a(x,v)\nabla z \cdot \nabla w . \tag{14.3.4}$$

The choice of z_n as a test function gives

$$\alpha \int_\Omega \frac{|\nabla z_n|^2}{(1+|v_n|)^\theta} \le \int_\Omega a(x,v_n)|\nabla z_n|^2 \le \beta \,\|\nabla z\|_{L^2(\Omega)}\,\|\nabla z_n\|_{L^2(\Omega)}$$

by assumption (14.1.1) and Hölder's inequality on the right-hand side. Since v_n belongs to B (see *Step I*), we obtain that $\|z_n\|_{H_0^1(\Omega)}$ is bounded, since

$$\frac{\alpha}{(1+R_1)^\theta}\|\nabla z_n\|^2_{L^2(\Omega)} \le \|z\|_{L^2(\Omega)}\|z_n\|_{L^2(\Omega)}\,.$$

Therefore there exist a subsequence and a function z_0 in $H_0^1(\Omega)$ such that $z_{n_k} \to z_0$ weakly in $H_0^1(\Omega)$ and a.e. in Ω. Fixing w in $H_0^1(\Omega)$, we have that

$$\int_\Omega a(x,v_{n_k})\nabla z_{n_k}\cdot\nabla w \to \int_\Omega a(x,v)\nabla z_0\cdot\nabla w$$

for every w in $H_0^1(\Omega)$, since $a(x,v_n)\nabla w \to a(x,v)\nabla w$ in $(L^2(\Omega))^N$ by Lebesgue's theorem. On the other hand, we deduce from (14.3.4) that

$$\int_\Omega a(x,v_{n_k})\nabla z_{n_k}\cdot\nabla w = \int_\Omega a(x,v)\nabla z\cdot\nabla w$$

for every w in $H_0^1(\Omega)$; therefore

$$\int_\Omega a(x,v)\nabla(z_0-z)\cdot\nabla w = 0$$

for every w in $H_0^1(\Omega)$. Choosing $z_0 - z$ as a test function and using that $a(x,s) > 0$, we deduce that $z_0 = z$. By Theorem 12.8, z_{n_k} are uniformly Hölder in $\overline{\Omega}$, since they are uniformly bounded and $\frac{\alpha}{(1+R_1)^\theta} \le a(x,v_{n_k}) \le \beta$ (that is, $a(x,v_n)$ are uniformly coercive and bounded). Arzelà-Ascoli theorem applies and gives the existence of a continuous function g in $\overline{\Omega}$ such that $z_{n_k} \to g$ in $C(\overline{\Omega})$. Since $z_{n_k} \to z$ a.e. in Ω, we infer that $z_{n_k} \to z$ in $L^\infty(\Omega)$. As the limit is independent of the subsequence, $z_n \to z$ in $L^\infty(\Omega)$.

Let us show that for every bounded $C \subset L^\infty(\Omega)$, then $\overline{S(C)}$ is compact. Let v_n be a sequence contained in the invariant set B. Then $\|z_n\|_{L^\infty(\Omega)} \le R_1$. As above we can say that z_n are uniformly Hölder. Arzelà-Ascoli theorem implies that S is completely continuous. □

14.4 The regularizing effects of a lower order term

In this section we are going to study the Dirichlet problem

$$\begin{cases} -\operatorname{div}(a(x,u)\nabla u) + u = f\,, & \text{in } \Omega\,,\\ u = 0\,, & \text{on } \partial\Omega\,. \end{cases} \tag{14.4.1}$$

We will show that the presence of the lower order term changes the nature of the existence results. Let $f_n = T_n(f)$ and define the following sequence of problems:

$$\begin{cases} -\mathrm{div}(a(x, T_n(u_n))\nabla u_n) + u_n = f_n\,, & \text{in} \quad \Omega\,, \\ u_n = 0\,, & \text{on} \quad \partial\Omega\,. \end{cases} \tag{14.4.2}$$

It is clear from Theorems 5.11 and 6.7 that there exists a $H_0^1(\Omega) \cap L^\infty(\Omega)$ solution to problem (14.4.2).

Lemma 14.11. *If u_n is a solution of* (14.4.2), *then* $\|u_n\|_{L^m(\Omega)} \le \|f_n\|_{L^m(\Omega)}$.

Proof. In the case where f belongs to $L^m(\Omega)$ with $m > 1$, we choose $|u_n|^{m-2}u_n$ as a test function in problems (14.4.2). By dropping the operator term and using the Hölder's inequality on the right-hand side, one has

$$\int |u_n|^m \le \int_\Omega f_n |u_n|^{m-2}u_n \le \|f_n\|_{L^m(\Omega)} \|u_n\|^{m-1}_{L^m(\Omega)}$$

which gives the result.

In the case where $f \in L^m(\Omega)$ with $m = 1$, we choose $\frac{T_k(u_n)}{k}$ as a test function in (14.4.2). Again, dropping the operator term of the left-hand side, one has

$$\int_\Omega u_n \frac{T_k(u_n)}{k} \le \int_\Omega |f|\,.$$

It is now sufficient to pass to the limit as $k \to 0$ and use Fatou's Lemma. □

Now we assume that (14.1.1) holds with $\theta > 1$. Let $f \in L^m(\Omega)$, with $m > \theta\frac{N}{2}$; we will prove the existence of bounded solutions.

Theorem 14.12. *If f belongs to $L^m(\Omega)$, with $m > \theta\frac{N}{2}$ and $\theta > 1$, then there exists a solution $u \in H_0^1(\Omega) \cap L^\infty(\Omega)$ of the boundary value problem* (14.4.1).

Proof. The use of $\left[\frac{(1+|u_n|)^{\theta-1}-(1+k)^{\theta-1}}{\theta-1}\right]^+ \mathrm{sgn}(u_n)$, $k > 0$, as a test function in (14.4.2) and Young's inequality imply that

$$\begin{aligned} \int_{A_k} a(x, T_n(u_n))\,|\nabla u_n|^2\,(1+|u_n|)^{\theta-2} + \int_{A_k} |u_n| \frac{(1+|u_n|)^{\theta-1} - (1+k)^{\theta-1}}{\theta-1} \\ \le \frac{1}{\theta-1} \int_{A_k} |f_n|\,[(1+|u_n|)^{\theta-1} - (1+k)^{\theta-1}] \\ \le \frac{C_\varepsilon}{\theta-1} \int_{A_k} |f_n|^\theta + \frac{\varepsilon}{\theta-1} \int_{A_k} [(1+|u_n|)^{\theta-1} - (1+k)^{\theta-1}]^{\frac{\theta}{\theta-1}}\,. \end{aligned}$$

Using the inequality

$$[(1+t)^{\theta-1} - (1+k)^{\theta-1}] \le \begin{cases} c_\theta t^{\theta-1}\,, & \forall\, t > k \ge 2^{\frac{\theta-2}{\theta-1}} - 1\,, c_\theta = 2^{\theta-2}\,, \text{if} \quad \theta \ge 2 \\ c_\theta t^{\theta-1}\,, & \forall\, t > k\,, c_\theta = 1\,, \text{if} \quad 1 < \theta < 2 \end{cases}$$

on the last term and assumption (14.1.1) on the first one, we obtain

$$\alpha \int_{A_k} \frac{|\nabla u_n|^2}{(1+|u_n|)^2} + \int_{A_k} |u_n| \frac{(1+|u_n|)^{\theta-1} - (1+k)^{\theta-1}}{\theta - 1}$$
$$\leq \frac{C_\varepsilon}{\theta-1} \int_{A_k} |f_n|^\theta + \frac{\varepsilon}{\theta-1} c_\theta^{\theta-1} \int_{A_k} [(1+|u_n|)^{\theta-1} - (1+k)^{\theta-1}]|u_n| .$$

We choose ε such that $\varepsilon(c_\theta)^{\theta-1} = \frac{1}{2}$. Then one obtains

$$\int_{A_k} \frac{|\nabla u_n|^2}{(1+|u_n|)^2} \leq C \int_{A_k} |f_n|^\theta ,$$

where C is a positive constant depending on θ and ε. By Hölder's inequality with exponent $\frac{m}{\theta}$, this implies that

$$\int_{A_k} \left| \nabla \log\left(\frac{1+|u_n|}{1+k}\right)\right|^2 \leq C \int_{A_k} |f_n|^\theta \leq C\|f\|^\theta_{L^m(\Omega)} [\operatorname{meas}(A_k)]^{1-\frac{\theta}{m}} .$$

Sobolev's inequality on the left-hand side implies

$$\left(\int_{A_k} [\log(1+|u_n|) - \log(1+k)]^{2^*}\right)^{\frac{2}{2^*}} \leq C\|f\|^\theta_{L^m(\Omega)} [\operatorname{meas}(A_k)]^{1-\frac{\theta}{m}} .$$

Now we set $\log(1+|u_n|) = v_n$ and $\log(1+k) = h$. Remark that $A_k = \{|u_n| > k\} = \{v_n > h\}$. Thus the last inequality gives

$$\left[\int_{\{v_n>h\}} |v_n - h|^{2^*}\right]^{\frac{2}{2^*}} \leq C\|f\|^\theta_{L^m(\Omega)} [\operatorname{meas}(\{v_n > h\})]^{1-\frac{\theta}{m}} .$$

Note that $(1-\frac{\theta}{m})\frac{2^*}{2} > 1$, since $m > \theta\frac{N}{2}$. Then Lemma 6.2 applies and gives that the sequence $\|v_n\|_{L^\infty(\Omega)} = \|\log(1+|u_n|)\|_{L^\infty(\Omega)}$ is bounded, that is, the sequence $\|u_n\|_{L^\infty(\Omega)}$ is bounded.

As in the proof of Theorem 14.2, u_n is a solution for n sufficiently large. □

Theorem 14.13. *Let $f \in L^m(\Omega)$, with $m > 1$.*

(1) *Assume that $m \geq \theta + 2$. Then there exists a distributional solution $u \in H_0^1(\Omega) \cap L^m(\Omega)$ to problem* (14.4.1).
(2) *Assume that $\frac{\theta+2}{2} < m < \theta + 2$. Then there exists a distributional solution $u \in L^m(\Omega) \cap W_0^{1,\frac{2m}{\theta+2}}(\Omega)$ to problem* (14.4.1).

Proof. Step I: Assume that $m \geq \theta + 2$. We use $[(1 + |u_n|)^{1+\theta} - 1]\mathrm{sgn}(u_n)$ as a test function in (14.4.2). Hölder's inequality on the right-hand side and assumption (14.1.1) on the left one imply

$$\int_\Omega |\nabla u_n|^2 \leq C\{1 + \|f_n\|_{L^m(\Omega)}\}\||u_n|^{(1+\theta)}\|_{L^{m'}(\Omega)},$$

where C is a positive constant depending on α and θ. By Lemma 14.11, the last term is bounded uniformly if $(1 + \theta)m' \leq m$, that is, $m \geq 2 + \theta$. Therefore, there exists a function $u \in H_0^1(\Omega)$ such that, up to a subsequence, $u_n \to u$ weakly in $H_0^1(\Omega)$, strongly in $L^1(\Omega)$ and a.e. in Ω.

It is easy to pass to the limit in problems (14.4.2) in order to prove that u is a solution to problem (14.4.1). For the first term of the left-hand side, it is sufficient to observe that $\nabla u_n \to \nabla u$ weakly in $L^2(\Omega)$ and by Lebesgue's theorem $a(x, T_n(u_n))\nabla\varphi \to a(x, u)\nabla\varphi$ in $(L^m(\Omega))^N$ for every m, due to assumption (14.1.1).

Step II: Assume that $\frac{\theta+2}{2} < m < \theta + 2$. If we choose $\varphi = [(1 + |u_n|)^{(m-1)} - 1]\mathrm{sgn}(u_n)$ as a test function in (14.4.2), we get, by the assumptions on a

$$\int_\Omega \frac{|\nabla u_n|^2}{(1 + |u_n|)^{\theta-m+2}} \leq C \int_\Omega |f|\,|u_n|^{m-1}.$$

Now, using Hölder's inequality on the right-hand side of the previous inequality and Lemma 14.11, we get

$$\int_\Omega \frac{|\nabla u_n|^2}{(1 + |u_n|)^{\theta-m+2}} \leq C\left[\int_\Omega |u_n|^m\right]^{1-\frac{1}{m}} \leq C\|f\|_{L^m(\Omega)}^{\frac{m}{m'}}, \quad \forall n \in \mathbb{N}, \tag{14.4.3}$$

where C depends on m and θ. On the other hand, let $\sigma < 2$. Writing $\int_\Omega |\nabla u_n|^\sigma$ as

$$\int_\Omega |\nabla u_n|^\sigma = \int_\Omega \frac{|\nabla u_n|^\sigma}{(1 + |u_n|)^{\frac{\sigma}{2}[\theta-m+2]}}(1 + |u_n|)^{\frac{\sigma}{2}[\theta-m+2]}$$

and using Hölder's inequality with exponent $\frac{2}{\sigma}$, we get

$$\int_\Omega |\nabla u_n|^\sigma \leq C\left(\int_\Omega (1 + |u_n|)^{\frac{\sigma}{2-\sigma}[\theta-m+2]}\right)^{1-\frac{\sigma}{2}},$$

where we have used (14.4.3). Due to Lemma 14.11, if $\frac{\sigma}{2-\sigma}[\theta - m + 2] = m$, that is, $\sigma = \frac{2m}{\theta+2}$, the right-hand side is uniformly bounded. Notice that $\sigma < 2$, by the assumptions on m and θ. Since $\frac{2m}{\theta+2} > 1$, the fact that the sequence $\int_\Omega |\nabla u_n|^{\frac{2m}{\theta+2}}$ is bounded, implies the existence of a function $u \in W_0^{1,\frac{2m}{\theta+2}}(\Omega)$ such that, up to a subsequence, $u_n \to u$ weakly in $W_0^{1,\frac{2m}{\theta+2}}(\Omega)$ and a.e. in Ω; moreover $u \in L^m(\Omega)$ by Lemma 14.11. One can prove that u is a distributional solution to problem (14.4.1), as in *Step I.* □

We are now going to study the case where $m = 2$, for $\theta = 2$ proving the existence of a $W_0^{1,1}(\Omega)$ solution to problem

$$\begin{cases} -\operatorname{div}(a(x,u)\nabla u) + u = f\,, & \text{in } \Omega\,, \\ u = 0\,, & \text{on } \partial\Omega\,. \end{cases} \tag{14.4.4}$$

We observe that the existence of a solution belonging to the nonreflexive space $W_0^{1,1}(\Omega)$ is quite unusual for an elliptic problem. However, also the Dirichlet problem

$$\begin{cases} -\operatorname{div}(|\nabla u|^{p-2}\nabla u) = f\,, & \text{in } \Omega\,, \\ u = 0\,, & \text{on } \partial\Omega\,, \end{cases}$$

with $1 < p < 2 - \frac{1}{N}$, and f belonging to $L^m(\Omega)$, $m = \frac{N}{(p-1)N+1}$, has a distributional solution $u \in W_0^{1,1}(\Omega)$, as proved in [14].

Theorem 14.14. *Let f be a $L^2(\Omega)$ function. Then there exists a distributional solution u in $W_0^{1,1}(\Omega) \cap L^2(\Omega)$ of (14.4.4), that is,*

$$\int_\Omega a(x,u)\nabla u \cdot \nabla\varphi + \int_\Omega u\,\varphi = \int_\Omega f\,\varphi, \quad \forall\,\varphi \in W_0^{1,\infty}(\Omega)\,. \tag{14.4.5}$$

If $f_n = T_n(f)$, by Theorem 14.12, there exists a bounded $H_0^1(\Omega)$ solution to the following problems:

$$\begin{cases} -\operatorname{div}(a(x,u_n)\nabla u_n) + u_n = f_n\,, & \text{in } \Omega\,, \\ u_n = 0\,, & \text{on } \partial\Omega\,. \end{cases} \tag{14.4.6}$$

We are going to use the previous approximating problems in order to prove Theorem 14.14. In the following, C will denote a positive constant depending on $\operatorname{meas}(\Omega)$.

Proof. Step I: We prove some *a priori* estimates on u_n. Let $k \ge 0$, $i > 0$, and let $\psi_{i,k}(s)$ be the function defined by

$$\psi_{i,k}(s) = \begin{cases} 0\,, & \text{if } 0 \le s \le k\,, \\ i(s-k)\,, & \text{if } k < s \le k + \frac{1}{i}\,, \\ 1\,, & \text{if } s > k + \frac{1}{i}\,, \\ \psi_{i,k}(s) = -\psi_{i,k}(-s)\,, & \text{if } s < 0. \end{cases}$$

Note that

$$\lim_{i \to +\infty} \psi_{i,k}(s) = \begin{cases} 1\,, & \text{if } s > k\,, \\ 0\,, & \text{if } |s| \le k\,, \\ -1\,, & \text{if } s < -k. \end{cases}$$

We will choose $|u_n|\,\psi_{i,k}(u_n)$ as a test function in (14.4.6); such a test function is admissible since u_n belongs to $H_0^1(\Omega)\cap L^\infty(\Omega)$ and $\psi_{i,k}(0)=0$. We obtain, by assumption (14.1.1)

$$\alpha\int_\Omega \frac{|\nabla u_n|^2}{(1+|u_n|)^2}|\psi_{i,k}(u_n)| + \int_\Omega u_n|u_n|\psi_{i,k}(u_n) \le \int_\Omega f_n|u_n|\psi_{i,k}(u_n)\,,$$

since $\psi'_{i,k}(s)\ge 0$. Using that $|f_n|\le|f|$, we have

$$\alpha\int_\Omega \frac{|\nabla u_n|^2}{(1+|u_n|)^2}\,|\psi_{i,k}(u_n)| + \int_\Omega u_n|u_n|\psi_{i,k}(u_n) \le \int_\Omega |f||u_n||\psi_{i,k}(u_n)|\,.$$

Letting i tend to infinity, we thus get, by Fatou's lemma on the left-hand side and by Lebesgue's theorem on the right-hand side,

$$\alpha\int_{A_k} \frac{|\nabla u_n|^2}{(1+|u_n|)^2} + \int_{A_k}|u_n|^2 \le \int_{A_k}|f|\,|u_n|\,. \tag{14.4.7}$$

Dropping the nonnegative first term in (14.4.7), and using Hölder's inequality on the right-hand side we obtain

$$\int_{A_k}|u_n|^2 \le \left[\int_{A_k}|f|^2\right]^{\frac12}\left[\int_{A_k}|u_n|^2\right]^{\frac12}.$$

Simplifying equal terms we thus have

$$\int_{A_k}|u_n|^2 \le \int_{A_k}|f|^2\,. \tag{14.4.8}$$

The previous inequality for $k=0$ reads

$$\int_\Omega|u_n|^2 \le \int_\Omega|f|^2\,, \tag{14.4.9}$$

so that u_n is bounded in $L^2(\Omega)$. This fact implies in particular that

$$\lim_{k\to+\infty}\operatorname{meas}(\{|u_n|\ge k\})=0\,,\quad \text{uniformly with respect to } n. \tag{14.4.10}$$

From (14.4.7), written for $k=0$, dropping the nonnegative second term, we have

$$\alpha\int_\Omega \frac{|\nabla u_n|^2}{(1+|u_n|)^2} \le \int_\Omega |f|\,|u_n|\,.$$

Hölder's inequality on the right-hand side then gives

$$\alpha\int_\Omega \frac{|\nabla u_n|^2}{(1+|u_n|)^2} \le \left[\int_\Omega|f|^2\right]^{\frac12}\left[\int_\Omega|u_n|^2\right]^{\frac12},$$

so that, by (14.4.9) we infer that

$$\alpha \int_\Omega \frac{|\nabla u_n|^2}{(1+|u_n|)^2} \le \int_\Omega |f|^2 . \tag{14.4.11}$$

Step II: We prove that, up to a subsequence, the sequence u_n strongly converges in $L^2(\Omega)$ to some function u belonging to $W_0^{1,1}(\Omega) \cap L^2(\Omega)$.

Starting from (14.4.11), we deduce that $v_n = \log(1+|u_n|)\mathrm{sgn}(u_n)$ is bounded in $H_0^1(\Omega)$. Therefore, up to subsequences, it converges to some function v weakly in $H_0^1(\Omega)$, strongly in $L^2(\Omega)$, and a.e. in Ω. If we define $u = [\mathrm{e}^{|v|} - 1]\mathrm{sgn}(v)$, then u_n converges a.e. to u in Ω. Let now E be a measurable subset of Ω; then

$$\int_E |u_n|^2 \le \int_{E\cap\{|u_n|\ge k\}} |u_n|^2 + \int_{E\cap\{|u_n|<k\}} |u_n|^2 \le \int_{A_k} |f|^2 + k^2 \mathrm{meas}(E) ,$$

where we have used (14.4.8) in the last passage. Due to (14.4.10), we may choose k large enough so that the first integral is small, uniformly with respect to n; once k is chosen, we may choose the measure of E small enough such that the second term is small. By Vitali's theorem, u_n strongly converges to u in $L^2(\Omega)$.

Step III: Let again E be a measurable subset of Ω, and let i be in $\{1, \dots, N\}$. Then, denoting by $\partial_i u_n$ the ith component of the distributional gradient of u_n, we have

$$\begin{aligned}
\int_E |\partial_i u_n| \le \int_E |\nabla u_n| &= \int_E \frac{|\nabla u_n|}{1+|u_n|} \\
&\le \left[\int_E \frac{|\nabla u_n|^2}{(1+|u_n|)^2}\right]^{\frac{1}{2}} \left[\int_E (1+|u_n|)^2\right]^{\frac{1}{2}} \\
&\le \left[\frac{1}{\alpha}\int_\Omega |f|^2\right]^{\frac{1}{2}} \left[2\mathrm{meas}(E) + 2\int_E |u_n|^2\right]^{\frac{1}{2}} ,
\end{aligned}$$

where we have used (14.4.11) in the last inequality. Since the sequence u_n is compact in $L^2(\Omega)$, we have that the sequence $\int_E \partial_i u_n \to 0$, as $\mathrm{meas}(E) \to 0$, uniformly with respect to n. Thus, by Dunford–Pettis theorem, and up to subsequences, there exists Y_i in $L^1(\Omega)$ such that $\partial_i u_n$ weakly converges to Y_i in $L^1(\Omega)$. Since $\partial_i u_n$ is the distributional derivative of u_n, we have, for every n in $\mathbb{N}$,

$$\int_\Omega \partial_i u_n \, \varphi = -\int_\Omega u_n \, \partial_i \varphi , \quad \forall \varphi \in C_0^\infty(\Omega) .$$

We now pass to the limit in the above identities, using that $\partial_i u_n$ weakly converges to Y_i in $L^1(\Omega)$, and that u_n strongly converges to u in $L^2(\Omega)$; we obtain

$$\int_\Omega Y_i \, \varphi = -\int_\Omega u \, \partial_i \varphi , \quad \forall \varphi \in C_0^\infty(\Omega) ,$$

which implies that $Y_i = \partial_i u$, and this result is true for every i. Since Y_i belongs to $L^1(\Omega)$ for every i, u belongs to $W_0^{1,1}(\Omega)$, as desired.

Step IV: We are going to prove that u is a solution to problem (14.4.4). Since $a(x,u_n)\nabla\varphi \to a(x,u)\nabla\varphi$ a.e. in Ω, by Egorov's theorem, for every $\delta > 0$ there exists a subset E_δ of Ω, with $\text{meas}(E_\delta) < \delta$, such that

$$\lim_{n\to+\infty} a(x,u_n)\nabla\varphi = a(x,u)\nabla\varphi \quad \text{uniformly in } \Omega\setminus E_\delta. \tag{14.4.12}$$

We now have

$$\begin{aligned}&\left|\int_\Omega a(x,u_n)\nabla u_n\cdot\nabla\varphi - \int_\Omega a(x,u)\nabla u\cdot\nabla\varphi\right| \\ &\qquad \le \left|\int_{\Omega\setminus E_\delta} a(x,u_n)\nabla u_n\cdot\nabla\varphi - a(x,u)\nabla u\cdot\nabla\varphi\right| + \beta\int_{E_\delta}|\nabla\varphi||\nabla u_n|.\end{aligned}$$

Using the equiintegrability of $|\nabla u_n|$ proved above, and the fact that $|\nabla u|$ belongs to $L^1(\Omega)$, we can choose δ such that the second term of the right-hand side is arbitrarily small, uniformly with respect to n. We then use (14.4.12) to choose n large enough so that the first term is arbitrarily small. □

We are going to prove the uniqueness of the solution obtained by approximation. Let $f \in L^2(\Omega)$, let f_n be a sequence of $L^\infty(\Omega)$ functions converging to f in $L^2(\Omega)$, with $|f_n| \le |f|$, and let u_n be a solution of (14.4.6). We have just proved the existence of a distributional solution u in $W_0^{1,1}(\Omega)\cap L^2(\Omega)$ to (14.4.4), such that, up to a subsequence,

$$\lim_{n\to+\infty} \|u_n - u\|_{W_0^{1,1}(\Omega)\cap L^2(\Omega)} = 0. \tag{14.4.13}$$

Now, let $g \in L^2(\Omega)$, let g_n be a sequence of $L^\infty(\Omega)$ functions converging to g in $L^{\frac{\theta+2}{2}}(\Omega)$, with $|g_n| \le |g|$, and let z_n in $H_0^1(\Omega)\cap L^\infty(\Omega)$ be a weak solution of

$$\begin{cases} -\text{div}(a(x,u_n)\nabla u_n) + z_n = g_n, & \text{in } \Omega, \\ z_n = 0, & \text{on } \partial\Omega. \end{cases} \tag{14.4.14}$$

Then, up to a subsequence, we can assume that

$$\lim_{n\to+\infty} \|z_n - z\|_{W_0^{1,1}(\Omega)\cap L^2(\Omega)} = 0, \tag{14.4.15}$$

where z in $W_0^{1,1}(\Omega)\cap L^2(\Omega)$ is a distributional solution of

$$\begin{cases} -\text{div}(a(x,z)\nabla z) + z = f, & \text{in } \Omega, \\ z = 0, & \text{on } \partial\Omega. \end{cases} \tag{14.4.16}$$

Our result, which will imply the uniqueness of the solution by approximation of (14.1.2), is the following.

Theorem 14.15. *Assume that $a = a(x, s)$ is differentiable with respect to the second variable, and $\left|\frac{\partial a}{\partial s}\right| \le C$, for some positive constant C. Assume that u_n and z_n are solutions of* (14.4.6) *and* (14.4.14)*, respectively, and that* (14.4.13) *and* (14.4.15) *hold true, with u and z solutions of* (14.1.2) *and* (14.4.16) *respectively. Then*

$$\int_\Omega |u - z| \le \int_\Omega |f - g|. \tag{14.4.17}$$

Moreover,

$$f \le g \text{ a.e. in } \Omega \quad \text{implies} \quad u \le z \text{ a.e. in } \Omega. \tag{14.4.18}$$

Proof. Subtracting (14.4.14) from (14.4.6) we obtain

$$-\operatorname{div}(a(x, u_n)\nabla u_n - a(x, z_n)\nabla z_n) + u_n - z_n = f_n - g_n.$$

Choosing $T_h(u_n - z_n)$ as a test function we have

$$\begin{aligned}\int_\Omega [a(x, u_n)\nabla u_n - a(x, z_n)\nabla z_n] \cdot \nabla T_h(u_n - z_n) + \int_\Omega (u_n - z_n)T_h(u_n - z_n)\\ = \int_\Omega (f_n - g_n)T_h(u_n - z_n).\end{aligned}$$

This equality can be written in an equivalent way as

$$\begin{aligned}&\int_\Omega a(x, u_n)[\nabla u_n - \nabla z_n] \cdot \nabla T_h(u_n - z_n) + \int_\Omega (u_n - z_n)T_h(u_n - z_n)\\ &\quad = \int_\Omega (f_n - g_n)T_h(u_n - z_n) - \int_\Omega [a(x, u_n) - a(x, z_n)]\nabla z_n \cdot \nabla T_h(u_n - z_n).\end{aligned}$$

The first term of the left-hand side is nonnegative, so that it can be dropped; using Lagrange's theorem on the last term of the right-hand side, we therefore have, since the absolute value of the derivative of a with respect to the second variable is bounded,

$$\int_\Omega (u_n - z_n)T_h(u_n - z_n) \le \int_\Omega (f_n - g_n)T_h(u_n - z_n) + Ch\int_\Omega |\nabla z_n||\nabla T_h(u_n - z_n)|.$$

Dividing by h we obtain

$$\int_\Omega (u_n - z_n)\frac{T_h(u_n - z_n)}{h} \le \int_\Omega |f_n - g_n|\frac{|T_h(u_n - z_n)|}{h} + C\int_\Omega |\nabla z_n||\nabla T_h(u_n - z_n)|.$$

Since, for every fixed n, u_n and z_n belong to $H_0^1(\Omega)$, the limit as h tends to zero gives

$$\int_\Omega |u_n - z_n| \le \int_\Omega |f_n - g_n|,$$

which then yields (14.4.17) passing to the limit.

The use of $T_h(u_n - z_n)^+$ as a test function and the same technique as above imply that

$$\int_\Omega (u_n - z_n)^+ \le \int_{\{u_n \ge z_n\}} (f_n - g_n)\,.$$

Hence, passing to the limit as n tends to infinity, we obtain, if we suppose that $f \le g$ a.e. in Ω,

$$\int_\Omega (u - z)^+ \le \int_{\{u \ge z\}} (f - g) \le 0\,,$$

so that (14.4.18) is proved. □

Due to (14.4.17), we can prove that problem (14.1.2) has a unique solution obtained by approximation.

Corollary 14.16. *There exists a unique solution obtained by approximation of* (14.1.2), *in the sense that the solution u in $W_0^{1,1}(\Omega) \cap L^2(\Omega)$ obtained as limit of the sequence u_n of solutions of* (14.4.6) *does not depend on the sequence f_n chosen to approximate the datum f in $L^2(\Omega)$.*

Our last result is a nonexistence result for solutions of (14.1.2) in the case where the source is a Dirac mass.

Theorem 14.17. *Let μ denote a Dirac mass concentrated on a point of Ω. Then there is no solution to*

$$\begin{cases} -\mathrm{div}\left(\dfrac{\nabla u}{(1 + |u|)^2}\right) + u = \mu\,, & \text{in } \ \Omega\,, \\ u = 0\,, & \text{on } \ \partial\Omega\,. \end{cases}$$

More precisely, if f_n is a sequence of nonnegative $L^\infty(\Omega)$ functions which converges to μ in the tight sense of measures, and if u_n is the sequence of solutions to (14.4.6), *then u_n tends to zero a.e. in Ω and*

$$\lim_{n\to+\infty} \int_\Omega u_n\,\varphi = \int_\Omega \varphi\, d\mu\,, \quad \forall\, \varphi \in W_0^{1,\infty}(\Omega)\,.$$

Remark 14.18. More in general the previous result can be proved for bounded Radon measures concentrated on a set of zero harmonic capacity (see [8]).

Proof. For every $\varepsilon > 0$ there exists a function ψ_ε in $C_0^\infty(\Omega)$ such that

$$0 \le \psi_\varepsilon \le 1\,, \quad \int_\Omega |\nabla \psi_\varepsilon|^2 \le \varepsilon\,, \quad \int_\Omega (1 - \psi_\varepsilon)\,d\mu \le \varepsilon\,.$$

Note that, as a consequence of the estimate on ψ_ε in $H_0^1(\Omega)$, and of the fact that $0 \le \psi_\varepsilon \le 1$, ψ_ε tends to zero in the weak* topology of $L^\infty(\Omega)$ as ε tends to zero.

If f_n is a sequence of nonnegative functions which converges to μ in the tight convergence of measures, then

$$0 \le \lim_{n\to+\infty} \int_\Omega f_n\,(1-\psi_\varepsilon) = \int_\Omega (1-\psi_\varepsilon)\,d\mu \le \varepsilon\,. \tag{14.4.19}$$

Let u_n be the nonnegative solution to the approximating problem (14.4.6). If we choose $1-(1+u_n)^{-1}$ as a test function in (14.4.6), we have, dropping the nonnegative lower order term,

$$\int_\Omega \left|\frac{\nabla u_n}{(1+u_n)^2}\right|^2 \le \int_\Omega f_n\,.$$

Therefore, up to a subsequence, there exist σ in $(L^2(\Omega))^N$ and ρ in $L^2(\Omega)$ such that

$$\lim_{n\to+\infty} \frac{\nabla u_n}{(1+u_n)^2} = \sigma\,, \qquad \lim_{n\to+\infty} \left|\frac{\nabla u_n}{(1+u_n)^2}\right| = \rho\,, \tag{14.4.20}$$

weakly in $(L^2(\Omega))^N$ and $L^2(\Omega)$ respectively.

The choice of $[1-(1+u_n)^{-1}](1-\psi_\varepsilon)$ as a test function in (14.4.6) gives

$$\begin{aligned}
&\int_\Omega \frac{|\nabla u_n|^2}{(1+u_n)^4}(1-\psi_\varepsilon) + \int_\Omega u_n[1-(1+u_n)^{-1}](1-\psi_\varepsilon)\\
&\quad = \int_\Omega f_n[1-(1+u_n)^{-1}](1-\psi_\varepsilon) + \int_\Omega \frac{\nabla u_n\cdot\nabla\psi_\varepsilon}{(1+u_n)^2}[1-(1+u_n)^{-1}]\\
&\quad \le \int_\Omega f_n(1-\psi_\varepsilon) + \int_\Omega \frac{\nabla u_n\cdot\nabla\psi_\varepsilon}{(1+u_n)^2}[1-(1+u_n)^{-1}]\,.
\end{aligned} \tag{14.4.21}$$

We study the right-hand side. For the first term, (14.4.19) implies that

$$\lim_{\varepsilon\to0^+}\lim_{n\to+\infty} \int_\Omega f_n\,(1-\psi_\varepsilon) = 0\,,$$

while for the second one, we have, using (14.4.20), and the boundedness of $[1-(1+u_n)^{-1}]$,

$$\lim_{n\to+\infty} \int_\Omega \frac{\nabla u_n\cdot\nabla\psi_\varepsilon}{(1+u_n)^2}[1-(1+u_n)^{-1}] = \int_\Omega \sigma\cdot\nabla\psi_\varepsilon[1-(1+u)^{-1}]\,.$$

Recalling that σ belongs to $(L^2(\Omega))^N$, that ψ_ε tends to zero in $H_0^1(\Omega)$, and using the boundedness $[1-(1+u)^{-1}]$, we have

$$\lim_{\varepsilon\to0^+}\lim_{n\to+\infty} \int_\Omega \frac{\nabla u_n\cdot\nabla\psi_\varepsilon}{(1+u_n)^2}[1-(1+u_n)^{-1}] = 0\,.$$

Therefore, since both terms of the left-hand side of (14.4.21) are nonnegative, we obtain

$$\lim_{\varepsilon\to0^+}\lim_{n\to+\infty} \int_\Omega \frac{|\nabla u_n|^2}{(1+u_n)^4}(1-\psi_\varepsilon) = 0\,.$$

The functional

$$v \in L^2(\Omega) \to \int_\Omega |v|^2(1-\psi_\varepsilon)$$

is weakly lower semicontinuous on $L^2(\Omega)$; this implies that

$$\int_\Omega |\rho|^2 = \lim_{\varepsilon\to 0^+}\int_\Omega |\rho|^2(1-\psi_\varepsilon) \le \lim_{\varepsilon\to 0^+}\lim_{n\to+\infty}\int_\Omega \left|\frac{\nabla u_n}{(1+u_n)^2}\right|^2(1-\psi_\varepsilon) = 0\,,$$

that is, $\rho = 0$. Thus, since

$$\frac{\nabla u_n}{(1+u_n)^2} = \nabla\Big(1-(1+u_n)^{-1}\Big)\,,$$

by (14.4.20) the sequence $1-(1+u_n)^{-1}$ weakly converges to zero in $H_0^1(\Omega)$, and so, up to a subsequence, it strongly converges to zero in $L^2(\Omega)$. Therefore u_n tends to zero a.e. in Ω. Since the limit does not depend on the subsequence, the whole sequence u_n tends to zero a.e. in Ω.

For Φ in $(L^2(\Omega))^N$, by (14.4.20), one has

$$\left|\int_\Omega \sigma\cdot\Phi\right| = \lim_{n\to+\infty}\left|\int_\Omega \frac{\nabla u_n}{(1+|u_n|)^2}\cdot\Phi\right| \le \int_\Omega \rho\,|\Phi| = 0\,,$$

which implies that $\sigma = 0$. Therefore, passing to the limit in (14.4.6), that is, in

$$\int_\Omega \frac{\nabla u_n\cdot\nabla\varphi}{(1+u_n)^2} + \int_\Omega u_n\,\varphi = \int_\Omega f_n\,\varphi\,,\quad \varphi\in W_0^{1,\infty}(\Omega)\,,$$

we get, since the first term tends to zero,

$$\lim_{n\to+\infty}\int_\Omega u_n\,\varphi = \int_\Omega \varphi\,d\mu\,,$$

for every $\varphi \in W_0^{1,\infty}(\Omega)$, as desired. □

Bibliography

[1] A. Ambrosetti and P. H. Rabinowitz. Dual variational methods in critical point theory and applications. *J. Functional Analysis* 14: 349–381, 1973.

[2] A. Ambrosetti and G. Prodi. *A primer of nonlinear analysis*, volume 34 of *Cambridge Studies in Advanced Mathematics*. Cambridge University Press, Cambridge, 1993.

[3] M. Artola. Sur une classe de problèmes paraboliques quasi-linéaires. *Boll. Unione Mat. Ital.* 5-B: 51–70, 1986.

[4] Ph. Bénilan, L. Boccardo, T. Gallouët, R. Gariepy, M. Pierre and J. L. Vázquez. An L^1-theory of existence and uniqueness of solutions of nonlinear elliptic equations. *Ann. Scuola Norm. Sup. Pisa Cl. Sci. (4)*, 22(2): 241–273, 1995.

[5] L. Boccardo. T-minima: an approach to minimization problems in L^1. *Ricerche Mat.* 29: 135–154, 2000.

[6] L. Boccardo. A contribution to the theory of quasilinear elliptic equations and application to the minimization of integral functionals. *Milan Journal of Mathematics* 79: 193–206, 2011.

[7] L. Boccardo and G. Croce. *Esistenza e regolarità di soluzioni per alcuni problemi ellittici, Quaderni dell'Unione Matematica Italiana*. Pitagora Editrice, Bologna, 2010.

[8] L. Boccardo, G. Croce and L. Orsina. Nonlinear degenerate elliptic problems with $W_0^{1,1}(\Omega)$ solutions, *Manuscripta Math.* 137: 419–439, 2012.

[9] L. Boccardo and B. Dacorogna. Monotonicity of certain differential operators in divergence form. *Manuscripta Math.* 64: 253–260, 1989.

[10] L. Boccardo, A. Dall'Aglio and L. Orsina. Existence and regularity results for some elliptic equations with degenerate coercivity. *Atti Sem. Mat. Fis. Univ. Modena* 46: 51–81, 1998.

[11] L. Boccardo, V. Ferone, N. Fusco and L. Orsina. Regularity of minimizing sequences for functionals of the calculus of variations via the Ekeland principle. *Differential Integral Equations* 12: 119–135, 1999.

[12] L. Boccardo and T. Gallouët. Nonlinear elliptic and parabolic equations involving measure data. *J. Funct. Anal.* 87(1): 149–169, 1989.

[13] L. Boccardo and T. Gallouët. Nonlinear elliptic equations with right hand side measures. *Comm. P.D.E.* 17: 641–655, 1992.

[14] L. Boccardo and T. Gallouët. $W_0^{1,1}(\Omega)$ solutions in some borderline cases of Calderon–Zygmund theory. *J. Differential Equations* 253: 2698–2714, 2012.

[15] L. Boccardo, T. Gallouët and F. Murat. Unicité de la solution pour des équations elliptiques non linéaires. *C. R. Acad. Sci. Paris* 315(11): 1159–1164, 1992.

[16] L. Boccardo, T. Gallouët and J. L. Vázquez. Nonlinear elliptic equations in $\mathbf{R}^N$ without growth restrictions on the data. *J. Differential Equations* 105(2): 334–363, 1993.

[17] L. Boccardo and D. Giachetti. Existence results via regularity for some nonlinear elliptic problems. *Comm. Partial Differential Equations* 14: 663–680, 1989.

[18] L. Boccardo and D. Giachetti. A nonlinear interpolation result with application to the summability of minima of some integral functionals. *Discrete Cont. Dynam. Syst.* 11: 31–42, 2009.

[19] L. Boccardo and H. Brezis. Some remarks on a class of elliptic equations with degenerate coercivity. *Boll. Unione Mat. Ital.* 6: 521–530, 2003.

[20] L. Boccardo, F. Murat and J. P. Puel. L^∞-estimate for nonlinear elliptic partial differential equations and application to an existence result. *SIAM J.Math.Anal.* 23(2): 326–333, 1992.

[21] H. Brezis. Équations et inéquations non linéaires dans les espaces vectoriels en dualité. *Ann. Inst. Fourier (Grenoble)* 18(1): 115–175, 1968.

[22] H. Brezis. *Functional analysis, Sobolev spaces and partial differential equations*. Universitext. Springer, New York, 2011.

[23] H. Brezis and W. Strauss. Semi-linear second-order elliptic equations in L^1. *J. Math. Soc. Japan* 25: 565–590, 1973.

[24] H. Brezis and F. H. Browder. Strongly nonlinear elliptic boundary problems. *Ann. Scuola Norm. Sup. Pisa Cl. Sci (4)* 5(3): 587–603, 1978.

[25] F. H. Browder. Nonlinear elliptic boundary value problems. *Bull. Amer. Math. Soc.* 69: 862–874, 1963.

[26] E. De Giorgi. Teoremi di semicontinuità nel calcolo delle variazioni. *Lezioni tenute all'Istituto Nazionale di Alta Matematica, Roma, 1968–69*, appunti redatti da U. Mosco, G. Troianiello, G. Vergara.

[27] B. Dacorogna. *Direct methods in the calculus of variations*, volume 78 of *Applied Mathematical Sciences*. Springer-Verlag, Berlin, 1989.

[28] I. Ekeland. Nonconvex minimization problems. *Bull. Amer. Math. Soc. (N.S.)* 1(3): 443–474, 1979.

[29] L. C. Evans and R. F. Gariepy. *Measure theory and fine properties of functions*. Studies in Advanced Mathematics. CRC Press, BOCA, 1992.

[30] D. Gilbarg and N. S. Trudinger. *Elliptic partial differential equations of second order*. Classics in Mathematics. Springer-Verlag, Berlin, 2001. Reprint of the 1998 edition.

[31] S. Kakutani. Topological properties of the unit sphere of a Hilbert space. *Proc. Imp. Acad. Tokyo* 19: 269–271, 1943.

[32] P. D. Lax and A. N. Milgram. Parabolic equations. *Contributions to the theory of partial differential equations*, Annals of Mathematics Studies 33: 167–190. Princeton University Press, Princeton, N. J., 1954.

[33] C. Leone. T-minima: some remarks about existence, uniqueness and stability results. *Asymptot. Anal.* 29: 273–282, 2002.

[34] J. Leray and J.-L. Lions. Quelques résulatats de Višik sur les problèmes elliptiques nonlinéaires par les méthodes de Minty-Browder. *Bull. Soc. Math. France* 93: 97–107, 1965.

[35] J. Milnor. Analytic proofs of the "hairy ball theorem" and the Brouwer fixed-point theorem. *Amer. Math. Monthly* 85(7): 521–524, 1978.

[36] C. L. Dolph. Nonlinear integral equations of the Hammerstein type. *Trans. Amer. Math. Soc.* 66: 289–307, 1949.

[37] A. Manes and A. M. Micheletti. *Un'estensione della teoria variazionale classica degli autovalori per operatori ellittici del secondo ordine. Bollettino U.M.I.* 4(7): 285–301, 1973.

[38] G. Mingione. The Calderón-Zygmund theory for elliptic problems with measure data. *Ann. Sc. Norm. Super. Pisa Cl. Sci.* 6(5): 195–261, 2007.

[39] G. J. Minty. On a "monotonicity" method for the solution of nonlinear equations in Banach spaces. *Proc. Nat. Acad. Sci. U.S.A.* 50: 1038–1041, 1963.

[40] L. Nirenberg. Remarks on strongly elliptic partial differential equations. *Comm. Pure Appl. Math.* 8: 649–675, 1955.

[41] R. S. Palais and S. Smale. A generalized Morse theory. *Bull. Amer. Math. Soc.* 70: 165–172, 1964.

[42] A. Porretta. Some remarks on the regularity of solutions for a class of elliptic equations with measure data. *Houston J. Math.* 26: 183–213, 2000.

[43] M. M. Porzio. A uniqueness result for monotone elliptic problems. *C. R. Acad. Sci. Paris Ser.1* 337: 313–316, 2003.

[44] A. Prignet. Remarks on existence and uniqueness of solutions of elliptic problems with right-hand side measures. *Rend. Mat. Appl. (7)* 15(3): 321–337, 1995.

[45] D. H. Sattinger. Monotone methods in nonlinear elliptic and parabolic boundary value problems. *Indiana Univ. Math. J.* 21: 979–1000, 1971/72.

[46] J. Schauder. Der fixpunktsatz in funktionalraumen. *Studia Math.* 2: 171–180, 1930.

[47] J. Serrin. Pathological solutions of elliptic differential equations. *Ann. Scuola Norm. Sup. Pisa Cl. Sci.* 18: 385–387, 1964.

[48] G. Stampacchia. Formes bilinéaires coercitives sur les ensembles convexes. *C. R. Acad. Sci. Paris* 258: 4413–4416, 1964.

[49] G. Stampacchia. *Équations elliptiques du second ordre à coefficients discontinus.* Séminaire de Mathématiques Supérieures, No. 16 (Été, 1965). Les Presses de l'Université de Montréal, Montreal, Que., 1966.

[50] G. Stampacchia. *Le problème de Dirichlet pour les équations elliptiques du second ordre à coefficients discontinus. Ann. Inst. Fourier (Grenoble)* 15(1): 189–257, 1965.

[51] G. Stampacchia. *Variational inequalities.* Theory and Applications of Monotone Operators (Proc. NATO Advanced Study Inst., Venice, 1968), pp 101–192. Edizioni Oderisi, Gubbio, 1969.

[52] N. S. Trudinger. Linear elliptic operators with measurable coefficients. *Ann. Scuola Norm. Sup. Pisa* 27(3): 265–308, 1973.

Index

De Gruyter Studies in Mathematics

Volume 55
Lucio Boccardo, Gisella Croce
Elliptic Partial Differential Equations, 2013
ISBN 978-3-11-031540-0, e-ISBN 978-3-11-031542-4, Set-ISBN 978-3-11-031543-1

Volume 54
Yasushi Ishikawa
Stochastic Calculus of Variations for Jump Processes, 2013
ISBN 978-3-11-028180-4, e-ISBN 978-3-11-028200-9, Set-ISBN 978-3-11-028201-6

Volume 53
Martin Schlichenmaier
Krichever–Novikov Type Algebras: Theory and Applications, 2014
ISBN 978-3-11-026517-0, e-ISBN 978-3-11-027964-1, Set-ISBN 978-3-11-028025-8

Volume 52
Miroslav Pavlovic
Function Classes on the Unit Disc: An Introduction, 2014
ISBN 978-3-11-028123-1, e-ISBN 978-3-11-028190-3, Set-ISBN 978-3-11-028191-0

Volume 51
Daniel Groves
An Invitation to Hyperbolic Groups, 2014
ISBN 978-3-11-026277-3, e-ISBN 978-3-11-026280-3, Set-ISBN 978-3-11-916681-2

Volume 50
Zoltán Sasvári
Multivariate Characteristic and Correlation Functions, 2013
ISBN 978-3-11-022398-9, e-ISBN 978-3-11-022399-6, Set-ISBN 978-3-11-174044-7

Volume 49
Mikhail I. Ostrovskii
Metric Embeddings: Bilipschitz and Coarse Embeddings into Banach Spaces, 2013
ISBN 978-3-11-026340-4, e-ISBN 978-3-11-026401-2, Set-ISBN 978-3-11-916622-5

Bei Fragen zur Produktsicherheit wenden Sie sich bitte an:
If you have any questions regarding product safety,
please contact:

Walter de Gruyter GmbH
Genthiner Straße 13
10785 Berlin
productsafety@degruyterbrill.com